The Handbook of Market Research for Life Science Companies

The Handbook of Market Research for Life Science Companies

Finding the Answers You Need to Understand Your Market

By
Jean-François Denault

CRC Press
Taylor & Francis Group
Boca Raton London New York

CRC Press is an imprint of the
Taylor & Francis Group, an **informa** business

CRC Press
Taylor & Francis Group
6000 Broken Sound Parkway NW, Suite 300
Boca Raton, FL 33487-2742

Printed on acid-free paper

International Standard Book Number-13: 978-1-138-71356-7 (Paperback)
International Standard Book Number-13: 978-1-138-71357-4 (Hardback)

Library of Congress Cataloging-in-Publication Data

Names: Denault, Jean-Francois, author.
Title: The handbook of market research for life science companies /
Jean-Francois Denault.
Description: Boca Raton, FL : CRC Press, [2017] | Includes bibliographical
references and index.
Identifiers: LCCN 2017005024| ISBN 9781138713567 (pbk. : alk. paper) | ISBN
9781315198606 (ebook)
Subjects: LCSH: Biotechnology industries--Research. | Marketing research.
Classification: LCC HD9999.B442 D463 2017 | DDC 660.6068/8--dc23
LC record available at https://lccn.loc.gov/2017005024

Visit the Taylor & Francis Web site at
http://www.taylorandfrancis.com

and the CRC Press Web site at
http://www.crcpress.com

Contents

Preface

Market research is a well-documented business activity. If you are looking for information on market research, you will find many well-written resources and books on the topic.

So why write a book specifically for start-up companies in life sciences?

Most of the time, life science start-ups are trailblazing. They develop new technologies with no defined market. They redefine markets by developing a technology that goes across multiple verticals. As these start-ups perpetually redefine science and technology, they constantly reshape and redefine the markets they exist in.

As such, a comprehensive understanding of the market is crucial. Potential investors and partners seek companies with an accurate read of their market, and pitching to them with a great technology but with a poor understanding of the market, and a poor grasp on how this technology can integrate the market, is ill-advised.

This book provides an introduction to market research for those who need a practical playbook and who are active in an emerging market. It is intended for

- Start-ups or small companies looking to define their markets and needing tools to better develop an understanding of their market

■ Established market research practitioners, looking to increase their knowledge of market research for the life sciences industry

To complete this book, we interviewed life science venture capitalists to get their perspective on market research, and to get their perspective on what's of specific interest and what grabs their attention. As such, I want to personally thank all the individuals who accepted to share some time to discuss the topic of my book and their insight. M. Ajit Singh, Managing Director and General Partner, Artiman Ventures, Caroline Stout, Investor at EcoR1 Capital, Nicola Urbani, Investment Director at Emerillon Capital and Elizabeth Douville, Partner at AmorChem Financial Inc., a heartfelt thank you for your insights and for giving the book an additional perspective.

I also want to thank my many clients who I have had the extreme pleasure of working with, and who graciously accepted to share some of the work we did together throughout the years. A huge thank you to

■ Dennis Leggett, CEO of Sympretek Inc.
■ Kathy May, CEO of ShippingEasy
■ Wesley Baker, CEO of Ancon Medical Inc.
■ Claude Leduc, CEO of MRM Proteomics
■ Perry Niro, CEO of Pharmed Canada

Finally, a profound thank you to my wife, Corinne, and my kids, Jaz and Chad, for their support throughout this project.

It is my hope that this book can help you on your journey to innovation.

Introduction to Market Research

Introduction

Exubera, an inhaled insulin drug, is a captivating example of an innovative product that might have benefited from stronger market research. It was foreshadowed as a great technical achievement, stabilizing the insulin molecule to make it bioavailable in dry powder form. Furthermore, it was the first insulin product that did not need to be injected, so there were high expectations: Pfizer estimated annual sales would total anywhere from $1 to $2 billion. But those sales never materialized. After nine months, sales of the product barely broke $12 million, and captured less than 1% of the market. It was soon removed from the market. In Pfizer's own words, "Pfizer has made this decision because too few patients are taking Exubera."*

---

* Dariman, T. 2007. Exubera inhaled insulin discontinued. Diabetes Self-Management. https://www.diabetesselfmanagement.com/blog/exubera-inhaled-insulin-discontinued/ (Accessed December 25, 2016).

What Happened?

Blame has been attributed to everything from product management issues to weak advertising, but one thing is sure: There was weak consumer interest for the product, even if it was one of the best clinical products at the time. The reasons for this lack of interest were numerous. For one, users were embarrassed to use the device (it looked like a marijuana bong, was clunky to carry around and to manipulate, and was anything but inconspicuous). Furthermore, the device was difficult to use (you had to carefully measure your dosage) and patients felt that the training was overly complex.

Like most other patients, individuals with diabetes prefer solutions that are easy to use and discreet. Diabetes patients often feel stigmatized: if you handle a syringe and a vial in a public setting for an insulin injection, you might have the impression that you look like a drug addict, not like somebody who has a chronic disease. Case in point, sensor-augmented insulin pumps (which combine the technology of an insulin pump with a continuous glucose monitoring sensor) are increasing in popularity. These devices are worn directly under clothing, giving diabetes patients the privacy and convenience they desire.

As such, by not accounting for how the consumer would feel when using the product, Pfizer unknowingly had two issues to address to increase user adoption. First, the training for the device had to be optimized for its complexity. Second, Pfizer had to find a way to attenuate the perceived stigma that made the device inherently incompatible with user expectations. It is possible that focus groups with end users or more in-depth interviews with initial users would have alleviated some of the issues that Pfizer faced, or at least given them the insight they seem to have lacked.

Companies in life sciences cannot be concerned uniquely with safety and efficacy; they have to make sure that they have an accurate read on their market and their end users.

Furthermore, they have to make sure that their product fills a real gap in the market, both from a health and a consumer perspective. Market size is not the only metric of potential success, but other elements, such as the competitive landscape, consumer perception and interests as well as emerging technologies, all play a role in evaluating a potential market.

The objective of this book is to fill the role of a handbook in market research for life sciences organizations, helping them organize their market research efforts as well as developing the tools they need to collect, analyze, model, and present their data.

Importance of Market Research

According to AcuPoll, a market research agency, 95% of new products introduced each year fail, resulting in massive losses.

What's more astonishing is that while 80% of brands are engaged in market research, 43% of bottom-performing brands are also engaged in market research activities.* Hence, not only top performers are engaging in market research, but many lower performers are also engaging in market research, but they are just not getting good results. There are many reasons for this: sometimes, poor planning around market research limits its effectiveness, sometimes data is being collected incorrectly, and sometimes the data is not actionable or is incorrectly analyzed.

It is not unusual for organizations or founders to believe that they already know their market. Perhaps the decision makers have been involved in the market for a long time. Perhaps they operated in parallel industries, and are making similar assumptions (what works in industry A works in industry B). Or maybe a key opinion leader in the organization is

---

* Glowa, T. 2015. Caution: How Market Researchers are contributing to Product Failure. Insights Association. http://www.marketingresearch.org/article/caution-how-market-researchers-are-contributing-product-failure (Accessed December 25, 2016).

shaping the opinion of other decision makers, silencing the critical elements necessary to challenge existing assumptions.

But markets are fluid environments. They change rapidly, and yesterday's certainty might already be inaccurate. Also, it is very difficult to convince stakeholders that you know a market based on assumptions. Bringing compelling arguments based on market research is the key to convincing people you have the right idea.

Hence, market research is done for two audiences. First, it is done for the organization itself to find the information it needs to make effective decisions, to validate market viability and opportunities, as well as to identify competitive threats and trends. *Market research is done when a lack of information will cost the organization more than the cost to acquire the information.*

Second, market research is done to convince stakeholders and demonstrate the validity of an organization's business plan elements (such as the business model, market entry strategy, product viability, and more). It is used to support (or disprove) internal assessments and beliefs. *The results of market research are a key element of a company's business case.*

On the importance of market research ...

You have to go beyond asking yourself if there is a need for your innovation from the patient's or user's point of view. You have to establish whether there is a market for your product/service. Furthermore, beyond asking yourself who the end-users are, it is important to identify how much will this product/service cost? How much will it save to healthcare systems and third-party payers? Who will pay for it? Market penetration is very complex, especially when you go global, as every country has its own dynamics. Understanding your market complexity and integrating the various strategies needed to maximize market penetration will be key in your commercialization success.

Nicola Urbani, Investment Director at Emerillon Capital

We break down market research into four components: the process of acquiring the information (which we will detail in Chapters 2 and 3), the analysis of acquired data (which is the topic of Chapter 4), building models for acquired data (Chapter 5), and the presentation of data in a way that is clear and concise (which is explored in Chapter 6).

Basic Market Research Concepts

Market research is full of dichotomies, and to better align expectations, it's best to understand the differences between various concepts such as primary and secondary research, quantitative and qualitative research, and inch-deep, mile-wide and deep-dive research.

Primary and Secondary Market Research

There are two types of research that you can engage in: primary and secondary research.

- *Primary market research* is a collection of activities that the researcher is engaged in to create data. This could be done through a web survey, a series of in-depth interviews, or leading focus groups for example. As such, it is data that did not exist until the researcher completed the market research activity, and is tailored specifically to his needs.* This will be explored in depth in Chapter 2.
- *Secondary market research* involves data that already exists that the researcher is collating. It could be data that he collects through a web search, or by aggregating news posts or blogs for example. The researcher collects and then transforms the data into something coherent and useful. This will be the topic of Chapter 3.

---

* Whenever a gender-specific term is used, it should be understood as referring to both genders, unless explicitly stated.

Quantitative and Qualitative Data

Data can be quantitative or qualitative.

- *Quantitative data* refers to data that can be measured and numbered. Counting the number of potential clients for a product, calculating the number of products or doses of a drug a consumer uses each day, or measuring the average distance a patient is willing to travel to visit a specialized clinic are all different types of quantitative data. People in life sciences are usually quite familiar with using quantitative data to quantify the technical aspects of products. Quantitative data in market research is often used to size markets and identify market segments and opportunities. As we will see later in this book, there are data collection tools that are better adapted for quantitative data. For example, surveys (both online and in person) are usually the best way to generate an important quantity of quantitative data.

- *Qualitative data* is data that is subjective and subject to interpretation. It can include anything from stories to words, observations, pictures, and even audio recordings. Some examples of qualitative data include personal reasons for preferences in consumer products, the impact of quality on customer purchasing patterns, or the impact of packaging color on purchase decisions. Data collected through interviews, focus groups, and observation is usually of a qualitative nature, but as we will see later, it is possible to codify this data to turn it into quantitative data.

People who write reports usually prefer quantitative data for presentations or for decision making as the data "feels more real," and is easier to understand for the intended recipients. Surveying a large number of individuals, compiling the data, and transforming it into a compelling presentation (usually pie charts and frequency tables) is easier and is perceived as "real data" by most. But qualitative data has its place, as it can often be used to "give color" or meaning to quantitative data.

A simple example is a project I did a few years back. My client, an Australian ad agency, was working on a promotional campaign for a big pharmaceutical company. To assist in preparing the right message, we had surveyed over three thousand consumers on their usage of painkillers. The quantitative data demonstrated what the preferred brands were, but without context, we couldn't understand why the number one preferred brand was the one my client perceived as cheapest and less efficient. It was only after analyzing the qualitative data in the survey relative to why consumers made these purchasing decisions that we were able to uncover patterns in decision making (the major categories of reasons that people gave centered on topics such as the family's choice, health reasons, routine, price/sale and advertising). To deepen our understanding, we monitored spontaneous online discussions relating to the brands. We then found passion around the preferred product due to fewer secondary effects, debates on home brands versus regular brands, and patterns in how consumers perceived competing brands. We found that the leading brand, while perceived as weaker, was recommended more often, while the second most recommended brand was perceived as being tougher on pain, yet harder on digestion.

In another project, a non-profit association client had requested a series of deep-dive interviews with its members in order to understand underlying trends and concerns. As I presented the results of the study at an annual meeting, I was questioned quite a few times about the lack of tables and graphs. While I strived to explain that this was a qualitative study, and not a quantitative one, members could not wrap their heads around the qualitative dimensions, and the feedback from the room was that "the sample size was too small to make a decision" and that there was "not enough data." While the client was very happy with the end results, and there were some high quality findings that could be turned into actionable results, the lack of tables and graphs left many members in the room uncomfortable, and we were unable to move forward with the recommendations.

These examples bring up the importance of the convergence of data: quantitative data seems "more real" and is easier to convey, but qualitative data is often useful to understand and contextualize it.

Miles-Wide versus Deep-Dive Research

When approaching a market research project, a researcher will have to decide if he's going for a "miles-wide" approach, or going for a "deep dive."

The inch-deep, mile-wide approach indicates an overview of a segment, industry, or competitive landscape. As such, the research project is deployed to gather information on as many data points as possible simultaneously. By its nature, it is very exploratory. By the end of the market research initiative, you will be able to share a high-level overview of the specific topic.

The deep-dive approach implies focusing exclusively on a specific pre-defined topic. When doing this type of project, the top trends, competitors, or issues have already been identified, and the researcher deploys his energy to researching each of these topics in a very specialized approach. This means that rather than interviewing generalists, he will be interviewing key opinion leaders and consulting specialized resources.

Many projects will be a convergence of these two types of research, starting with the inch-deep, mile-wide approach, and concluding with a deep dive on the most interesting targets. Remember that it is important to align expectations with available resources.

A Word on Life Sciences

Life science is a broad term that is used in many different situations. In this book, life science includes all industries that have an impact on health (be it human or animal). As such, we are including pharmaceuticals, biotechnology, medical technologies, and nutraceuticals. We also include healthcare, but mostly as it pertains to how the four previous industries impact the healthcare sector.

This book will explore methodologies that apply to these sectors of activities. In an effort to englobe the most information, we might mention in passing other less relevant techniques, but our goal here is to zero in on the methods that might be used by start-ups and growing companies in life sciences, rather than be an all-inclusive repertoire of all that is market research.

As such, whenever possible, life sciences examples (ranging from the hypothetical to past projects I have worked on) will be used to illustrate the different concepts. By doing so, it becomes easier for you, the reader, to apply these methodologies to your own organization and projects.

Author

 **Jean-François Denault** has been work-
ing with innovators and entrepreneurs in
life sciences as a professional consultant
for over fifteen years. Through the years,
he has worked with over 40 different
clients in life sciences (including larger
companies such as J&J, Denka Seiken, and
Chemo Group). His clients are located throughout the world,
having completed projects with clients in over 25 different
countries.

Jean-François specializes in the life sciences segment. As
such, he has completed projects related to pharmaceuticals,
biotechnology, medical devices, nutraceuticals, and healthcare. Most of his projects have been in the market research,
marketing strategy, and competitive intelligence space. He
possesses a graduate degree in management consulting, an
executive MBA, and a graduate degree in organizational
communication.

He is a member of the editorial boards of the *Journal of
Brand Strategy* and the *Journal of Digital & Social Media
Marketing*, and has written a half-dozen articles for various
publications. He is on the advisory boards of several start-ups
(including Marshall Hydrothermal and JustBIO), is a strategic
advisor for Pharmed Canada (a non-profit organization

dedicated to the development of Canadian life science manu-
facturers and subcontractors) as well being a member of
Pharmed Canada's CMO-CDMO Strategic Committee and The
Laval Biotech City Life Science Advisory Committee. He is
also an active member of his community, having served on
multiple non-profit boards. Currently, he is the president of
the Lanaudière Alzheimer Society, a role he has been fulfilling
since 2012.

He is based out of Montreal, Canada, and can be reached at
jf.denault@impacts.ca.

Chapter 1

Market Research Basics

1.1 Introduction to Market Research Process

Coherent and valuable market research follows a systemized approach. As such, like many other processes in life, it all starts with building a plan. While it is tempting to jump directly into "market research" and start collecting data, a detailed market research plan ensures that the data collected will be consistent and useful. As we will see in this chapter, market research must be planned beforehand to ensure consistency in the data that is collected, as well as to formalize the end point.

Market research without a coherent plan runs the risk of shifting midway through the process, or even having to start over. Also, a carefully planned study ensures that all stakeholders have the same vision of why the study is being done, and what it is trying to discover.

Consider the following example. Many years ago, an advertising agency I was working with was developing continuing medical education (CME) content for a big pharmaceutical company. I was tasked with surveying doctors on their interest in participating in online learning CME courses. The client pushed this project forward as he felt it was urgent, and we sent out a survey questionnaire to over a thousand doctors. As results started to trickle in, we realized that some critical data was not being collected. This happened because in the rush to get the survey out, expectations were not aligned between all the stakeholders involved in the survey. We had to stop the survey, redesign the survey questionnaire, and resend it to participants. On top of looking unprofessional, it was a loss of time, resources, and credibility.

A carefully crafted coherent plan enables the market researcher to plan from the beginning to the end, and ensures time will be used more efficiently.

On the importance of doing your own market research ...

When we see the entrepreneur presenting a project with his own preliminary market research, it is a strong indication that he is proactively working to understand how his product fits in the market. He is moving beyond the technical applications of his innovation, towards its integration into the market, and it shows that he is determined to understand the issues beyond the lab. That is professionalism we favorably acknowledge.

Elizabeth Douville, Partner, AmorChem Financial Inc.

1.2 Market Research Process

Before describing the market research process, I want to stress that there is no one unique market research process. Some researchers break down the process into 4 steps, while others use over 11 steps. Each researcher has their own approach and way of subdividing the process. I use a simplified process that consists of four steps (Figure 1.1).

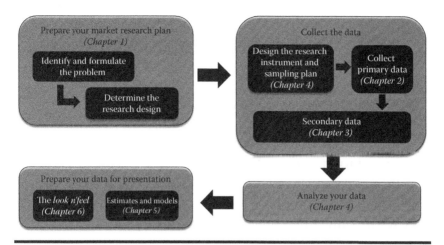

Figure 1.1   Market research process.

1.2.1 Prepare Your Market Research Plan

Preparing your market research plan breaks down into two distinct steps: (1) Identifying and formulating the problem, and (2) determining the research design.

a. *Identify and formulate the problem*: The first step of the market research process is to identify and formulate the problem (or the opportunity). By formally defining the problem, the market researcher will focus his research effectively, ensuring that all participants share the same vision and objectives for the project. As such, the problem identification step will usually involve discussions with decision makers, a review of secondary data, and conversations with key opinion leaders.

 The topic of research is usually defined in a few words. For example, it could be to identify emerging market opportunities for a new technology, the size and segment of the current market, or developing a customer profile.

b. *Determine the research design*: The next step is to determine the research design. It is the approach that you will use to collect your data, and will guide you in choosing the specific methods you will use to collect the information you need. Some key questions you will answer at this step are

■ Which method(s) will I use to collect data?
■ How will I connect with my data sample? Who will I need to connect with? How can I connect with them? Will I need to incentivize them? How?
■ Which data collection tools will I use (telephone, in-person, Internet)?
■ What is my total budget (both monetary and timewise)?

When determining your research approach, there are three types of research design you can choose. The three classifications are exploratory research, descriptive research, and causal research.

1.2.1.1 Exploratory Research

Exploratory market research is akin to basic research in science. It is done to better understand a phenomenon, or when the existing knowledge that the organization possesses is too vague. It helps the organization gain some broad insight and to learn more, enabling it to gain familiarity on a topic and to conduct more precise research. Usually, exploratory research is qualitative in nature, and uses techniques such as secondary research, focus groups, and interviews.

To illustrate this, let me share an example of exploratory research I did a few years ago. A client was interested in expanding into a new geographic market, but did not know which specific technologies were used in the targeted territory. To find out, I researched the topic online (secondary research) and contacted a few key opinion leaders that have experience in the region (interviews). After this project, the client did not have a formal market appraisal, but he did have an appreciation of which technologies competed on the market, which companies were present, and some key trends that would help him decide if he wanted to explore this market further.

1.2.1.2 Descriptive Research

Descriptive research is much more detailed and generates more granular data for the organization. This can only be done if the market researcher already has a good appreciation of the market, and can properly define his research needs. Quantitative data applies best for descriptive research and is mostly collected through surveys, although secondary research into specialized databases is also possible.

Continuing with our earlier example, we had identified that three technologies were present in the target market. At this point, the client was interested in knowing the market share of each product. We conducted a survey, sending it to relevant users. Once we collected sufficient data, we had an accurate quantitative appreciation of the market.

1.2.1.3 Causal Research

Causal research is the most specific type of research, and is usually done if sufficient data has already been collected. It is done to find specific explanations for specific issues. Most of the research will be qualitative, and the most popular methods of collecting data are observation, experimentation, and in-depth interviews.

Concluding our example, I had determined that a specific medical device had over 70% of the market share in the territory he wanted to expand into. My client's new research problem was finding out why this product had such a dominant position. To find out, he had different options. He could observe end users using the device, and compare that to end users using other devices. He could purchase each of the three devices on the market, use them, or have them used by an independent third party. Or he could do in-depth interviews of end users, focusing exclusively on how and why they use it to gain insight.

1.2.2 Data Collection Step

The data collection phase splits into distinct steps. First, there is the design step where you will design your sampling plan and your tools. This is followed by the collection of data.

1.2.2.1 Design Sampling Plan

The sampling plan is the detailed framework of who will be contacted and what the expected sample size is. The sample size is crucial for the validity of the information you collect. That's why calculating it beforehand is essential. There are two types of samples that a researcher can use: probabilistic and non-probabilistic.

Non-probabilistic samples are those where the participants are not selected at random. If you are consulting a defined group of experts, there is no need to build a survey group at random. Some non-probabilistic samples include:

- *Convenience sample*: Participants are selected because they are available and willing to participate. For example, if a researcher's company has a kiosk at a trade show, collecting data from participants at the trade show would be a convenience sample.
- *Quota sample*: If a researcher has decided beforehand that there are some minimum population requirements needed in his sample, he is building a quota sample. For example, he could decide that at least 50% of respondents should be women, or that 40% of respondents should be within a certain geographical area.
- *Snowball sample*: This is used in situations where participants are very specialized and hard to reach. Hence, KOLs are contacted and asked for contacts to populate

the survey. As participants take part in the survey, they invite others and so on. It can be necessary to start the recruitment process again once the sample is no longer supplying new participant leads.

■ *Voluntary sample*: When contacting a closed population, the sample is the participants who elect to participate. For example, members of an association who decide to participate in an association's annual survey serve as the basis of a voluntary sample.

As for probability sampling, it is a form of sampling that uses some form of random selection. If you are using *simple random sampling*, you select a number of participants out of a total population of users who each have an equal chance of being selected. *Stratified sampling* implies dividing your population into homogenous subgroups, then taking a simple random sample from each, while *systematic random sampling* implies selecting individuals of a population according to a random starting point and a fixed interval (e.g., surveying every three houses on a street).

A question that's often asked when doing a project is "How many people should we contact?" Sample size is not a standard number we can describe and there are no rules of thumb. That's why before doing descriptive research, it is necessary to do some exploratory research. This enables the market researcher to find an approximate market size in terms of users/potential clients, and he can then use a basic statistical model to find a relevant sample size. As such, once the number of current users is estimated, we can use an online sample size calculator to determine a valid sample size.

Of course, available budget will have an impact on the sample size. It might be nice to interview 500 people individually to get a scientifically valid sample, but time- and monetary-wise, such an endeavor would be cost-prohibitive. As such, sometimes compromises must be made. A researcher

Table 1.1 Example of a Sampling Plan

Parameters to be measured	User adoption of technology by radiologists in U.S. hospitals
Data to be collected	Factors impacting user adoption
Data collection method to be used	Surveys
Where will the data be collected	Online, using third-party online survey lists
Time line	Week #1: 1st invitation to participate Week #2: Reminder Week #3: Final reminder
Sample size	There are an estimated 38,000 radiologists working in hospitals in the United States To obtain a valid sample, we need to have responses from at least 384 individuals

might decide to interview only 20 people, which would already give him a wealth of information. In fact, a recent article mentioned that about 30 interviews were necessary if the researcher's objective was to generate subgroups during analysis.* Intuitively, once the information you get from interviewees starts to repeat itself, you will know that you have attained a sufficient sample size. A typical sampling plan is shown in Table 1.1.

1.2.2.2 Design Your Tool

Once the research question has been designed, and the methodology decided on, it is time to design the research tool. For example, if you have decided to do interviews, you will have

---

* Duncan B. 2015. *Can Qualitative Research Be Used to Detect Segments? Think RGA.* http://thinkrga.com/2015/10/sample-size-for-detecting-segments-in-qualitative-research/ (Accessed 29 January, 2017).

to design an interview guide. Building an interview guide ensures consistency between each interview, between each interviewer, as well being a useful tool for remembering topics during the interviews.

Ideally, you should test your research instrument before using it at large. Test your interview guide with a few potential participants: you might find that some questions are redundant, that some questions are missing, and that some questions are misunderstood by your target audience. It is much more cost-effective to find this out at this stage rather than at the data analysis stage.

The two main types of tools a researcher can prepare when doing market research are questionnaires and a research guide (which includes variants such as discussion and observation guides).

A questionnaire is a list of questions that a researcher prepares to collect data from a statistically significant number of subjects. It is mostly used in surveys (online or in person), telephone, and mailing interviews. Most of the time, the market researcher will not be administrating the survey as it will be done by a third party (such as an online web tool, or a paid resource). As such, the researcher has to make sure that the questions are sufficiently clear, that misinterpretations are unlikely, and that the collection is standardized across multiple interviewers. As for the discussion guide, it is mostly used in focus groups and in-depth interviews, and includes mainly open-ended question that guide the conversation.

More information on building these tools is available in Chapter 4.

1.2.2.3 Collecting Data

This is the moment when you go down to the trenches and put your plan into action. You start recruiting and doing interviews, you send out your survey, or you start searching the Internet for the information you need. This could be done by you, or you might hire a firm to do some (or all) of the data collection. We will go over this step in detail in Chapters 4 and 5, but the important thing to remember is that this step is often the most time-consuming step of the market research process.

1.2.4 Analyze Your Data

Once you have reached the end point of your data collection, it is time to start analyzing data.

If you have quantitative data, you might use spreadsheet software (such as Excel) or a statistical software package (such as SPSS or JMP). You will build tables and graphs correlated with the demographic variables (age, geography, etc.). You start looking for trends to find the story that you will be sharing with stakeholders.

If your data is qualitative, the first step is the transcription and codification of data. Codification enables the market researcher to identify patterns in responses. For example, you might start to formulate broad categories around responses, creating buckets for the data to make sense. For some projects, it might be useful for two individuals to independently go over the data, and then compare results to try to reduce bias. We will go over various methods to make sense of data in Chapter 6.

1.2.5 Prepare Your Data for Presentation

Once you have interpreted your data, and transformed the information from raw data into a coherent story, it is time to share it with stakeholders, investors, and potential partners. You might want to spend some time presenting the methodology used to find the data to showcase and reinforce its validity, but the bulk of the presentation will be on results, as well as the related recommendations and insights that were generated following your research.

This section is so important that two whole chapters will be dedicated to it. One chapter will go over different ways to present data, and how to classify it in a coherent way to generate additional insights. This is the topic of Chapter 5. Chapter 6 will be dedicated to the "look n' feel," so we will be showcasing how to present the data in a consistent and punchy manner as well as learning what not to do in your presentation.

1.3 Case Study: Market Research Process in Action

To illustrate the process, here is a mini case study based on a project I did a few years ago (see Table 1.2). At the time, a client had developed an innovative insulation solution that could be used in multiple markets, but was not familiar with the shipping and transportation of medical products. He wanted to assess the feasibility of targeting the medical shipping market as a short-term market for his product, and he wanted to understand the main trends and decision factors when selecting a shipping solution. For a sample market research plan (See Table 1.2).

Table 1.2   Sample Market Research Plan

1. *Identify and formulate the problem*	
• How do shipping managers select the shipping packages they use? (What is the process?) • What are the main decision factors?	
2. *Determine the research design*	
• Exploratory research to gain a high-level understanding of how shipping managers select their shipping packaging and what are the main decision points	
3. *Design the research instrument and sampling plan*	
Parameters to be measured	Decision factors when purchasing shipping products
Data to be collected	The top decision factors
Data collection method to be used	In-depth interviews
How will the data be collected	Phone
Time line	3 weeks for recruitment and interviews 1 week for analysis and presentation
Sample size and characterization	10 shipping managers or decision makers working in shipping in the pharmaceutical/biotechnology industry
Research instrument	Discussion guide
4. *Collect the data*	
• Perform 10 in-depth interviews by phone	
5. *Analyze the data*	
• Codify and analyze interviews	
6. *Present the findings*	
• Presentation online (Webex) to client	

Chapter 2

Primary Research

Primary data is information that is generated directly by the market researcher to answer his research question. For example, when he is doing interviews, online surveys, or making observations, he is gathering primary data.

Generally, primary research is costlier to generate (both in terms of time and resources), but it is customized for the researcher's needs. If he has correctly designed his tools, he should be able to solve his research problems. Also, the data he collects is proprietary, so it belongs to the organization exclusively, becoming a competitive advantage.

This chapter is divided into two distinct sections. The first section deals specifically with the data collection framework (i.e., the questionnaire or discussion guide), while the second section deals with the information collection activities specifically, such as in-depth interviews, focus groups, surveys, observation, mystery shops, and Delphi groups, from an implementation perspective and provides examples and information on how these activities are used by life science organizations.

2.1 Importance of Preparing a Market Research Tool

The preparation of your market research tools is an important part of the data collection process as it ensures the quality of the data collected by making sure that

a. The data is collected in a *consistent manner*. To be able to compile data from different investigators, or from different time periods, it has to be collected in a consistent manner. The questions have to remain the same from one sample to the next. It is problematic to compile data if some respondents answered questions that were phrased differently.

   For example, a few years back, I was brought in to a project to analyze some primary data a client had collected, as he was unable to build consistent models. A web survey had been posted on two different websites that belonged to two different subsidiaries. Each subsidiary had designed a different autonomous web survey and had collected the data independently. After a careful review, I noticed some subtle yet significant differences between the questions in the two surveys. In an effort to "improve" demographic data, questions and answers were changed in one of the two surveys. The changes were significant enough that the online survey was not collecting the data the same way in the two locations. As such, it was impossible to merge the data from the two data sources, and the data was impossible to tabulate correctly until some of the data was recoded in one of the surveys.

b. The researcher does not forget any questions and covers every topic: This is more important for voice interviews than written data collection activities. During an interview, it is easy to get wrapped up in the conversation and forget to ask some questions on a specific topic, skipping some questions. While it is occasionally possible to go back and ask the person interviewed, the dynamic created in the initial interview is lost.

2.2 Designing a Data Collection Tool: Step by Step

Building your data collection tool is an important step. If possible, multiple individuals should be implicated at various stages to ensure that all the information needed is collected. We propose here four simple steps to develop your data collection tool.

2.2.1 Step One: Define the Context

As discussed earlier, the first step in doing market research is to define the information required, the target respondents and data collection tool you will use.

- *Defining the information required* helps the researcher focus his questions. While he might have defined his needs from a broad perspective, it might be useful to do a little secondary research first to gain some additional insight and help clarify what information really needs to be created versus what already exists.
- Next, identify the *target respondents*. Defining them clearly helps conceptualize the knowledge level of the respondent and their technical expertise, which influences what data collection tool you will use and the way the questions are going to be prepared.
- Finally, the choice of the *data collection tool* will dictate how the questions will be asked and worded. For example, a questionnaire might be self-administered (the participant is alone when answering the questions) or assisted (with an interviewer asking the questions,

clarifying topics if needed). If it is self-administered, you should take into account that the user will have to interpret the questions and answers by himself, and more detailed questions might be necessary.

2.2.2 Step Two: Build Your Question Bank

A brainstorming session with other interested parties is a great way to start building your question bank. During this creativity phase, the objective is to identify all the topics you want covered by your market research activity, so as to generate as many questions you think you will need to ask, and try to cover as much ground as possible. You can weed out duplicates and unnecessary questions when you build your data collection tool. We go into more detail on how to design questions in Section 2.3.

2.2.3 Step Three: Build Your Data Collection Tool

Once you believe you have generated enough questions to cover all the ground you need covered, it is time to place your questions in a meaningful order.

a. *Beginning of the questionnaire*
■ Start your data collection tool with a brief statement explaining why you are collecting the information and, if applicable, stating that the information is entirely confidential and that only consolidated results will be published.
   – Participants might be hesitant to answer questions and share sensible opinions if they don't know what the information is going to be used for. Clearly stating why you are collecting the data will contextualize the market research activity for the participants, and increase participation.
   – Reminding users at the start of the data collection tool that the information will not be shared at large

increases participation and ensures the participants share more truthfully, while mentioning that it will only be shared in a consolidated form makes it more believable for participants.

■ Include warm-up questions: Start your question guide with a few simple questions to engage the participants, and ensure that they feel comfortable in the process. If you're using a web-based platform, including a few easy warm-up questions lets the user gain familiarity with the online survey platform. Some simple warm-up questions could include questions about the person's job title and function, for example, or his relationship to the topic covered in the data collection tool.

b. *Middle of the survey*

■ Put your most important questions in the middle of your data collection tool. Do not put the most important questions at the end of the question guide, as you do not want your participants suffering from participant fatigue when they get to them as they may botch or skip them to finish the survey.

   – Place the most important items in the first half of the questionnaire to increase the response rate on the important topics: if you are short on completed surveys, you might be able to use partially completed surveys to fill the gaps.

■ Make sure the questions flow from one to another, and don't force the participants to go back and forth on topics. If you ask a series of questions on how they use a medical device, try to group these questions together to keep participants in the same mind-set.

■ When building your data collection tool, be critical of your questions. Ask yourself, "Is this question really useful?" Is it "need to know" or "nice to know"? Participant attrition occurs when data collection is too long, so make sure you only keep the important questions you need answered.

■ Include some variety in the type of questions you ask. Too many open-ended or closed-ended questions clustered together might bore participants. Instead, mix the type of questions you ask to sustain user participation.

c. *End of survey*

■ Remember that as the participant nears the end of the data collection tool, he might become increasingly indifferent, giving careless or half-thought-out answers (as he's just trying to reach the end of the survey to get his incentive).

■ Questions that are very sensitive in nature can be included at the end of the survey to avoid losing participants before other important information is collected.

– For example, asking a question on a sensitive topic (e.g., assisted suicide, abortion) might deject participants. Having the question at the end lets you use the other data they provided before dropping out.

■ At the end of the survey, remember to include questions related to demographics: This will ensure that you can compare the answers between segments, as well as correlate answers and identify trends.

– Some researchers prefer putting demographic questions at the beginning of the survey instead, using them as warm-up questions. That is acceptable and especially recommended if you have a few sensible questions at the end of your questionnaire, or feel that participant attrition will be high.

2.2.4 Step Four: Validation

Before deploying your data collection tool at large, test it with a sample of your target population. If your objective is to interview end users, try your questionnaire beforehand on a few end users to identify any gaps or misunderstandings your participants may have. This helps ensure that the questions are clearly asked, and that the data you collect is what you need as well as being easy to compile and interpret. It is also useful to make sure that the questions are answered consistently across different individuals. Finally, it is useful to validate your survey so that you can remove questions that are not really needed, or that are redundant.

Pre-testing your data collection tool also lets you validate its length. If you see participant fatigue during the testing phase, consider removing some questions, or adjusting them from open- to closed-ended. As a rule of thumb, an online survey should take about 20 minutes to fill out (unless you are offering an incentive or you have an especially captive audience) while an in-depth interview could reasonably last 30–45 minutes and a focus group could last between 60 and 90 minutes. Of course, if you are targeting participants that are very busy, a shorter tool is needed: a project targeting chief executive officers (CEOs) should include only four to five key questions to ensure a maximum participation rate.

After testing, you should be ready to start data collection.

2.3 Formulating Questions

In this section, we will go over some of the finer points relating to formulating questions. Take note that the golden rule of building questions is to *keep your questions simple*. The more complex the question is, the more likely the person answering it will either skip it, not answer truthfully, or just answer incorrectly as they aren't able to understand what you are trying to ask.

This section will deal with closed-ended questions and open-ended questions. As we discussed earlier, your market research can have an exploratory objective, a descriptive objective, and a causal objective. Some research models generally lend themselves to open-ended questions (exploratory and causal research mostly), while some others lend themselves to closed-ended questions (mostly descriptive research). Following some tips on writing these types of questions, we will go over projective and choice modeling questions, discuss the use of question banks and mindful surveys in helping you prepare your questions, and close with a few things to watch for.

2.3.1 Closed-Ended Questions

There are three types of closed-ended questions: nominal, ordinal, and interval questions.

Nominal questions are those where the choices are presented to participants across categories but in no particular order. Asking users which color they prefer, or which brand of painkiller they use is a nominal question.

> An important side note here is that the order that options are presented to a respondent can influence their answer choices. This is called the Primacy Effect: in paper and internet surveys, respondents tend to pick the first choice of a list, rather than reading through every available option. If you are using an online survey tool, use a random sorting option to attenuate the Primacy Effect.

Ordinal questions are used when there are different answer categories, with an importance ranking across possible answers. For example, a market researcher may ask to rank a product across a scale of slow, medium, and fast.

> It is important to make sure that there is no overlap or confusing terms across categories. For example, a scale of "small, average, large, big" possesses several issues: What is average? What is the difference between large and big? A much better scale would simply be "small, medium and large."

An *interval question* is one where spacing across categories is even. For example, asking users how much they like something on a scale of one to seven, where 1 is "not at all" and 7 is "the most" is an interval question. A good interval scale will usually have five to seven points and will include a middle category to produce better data, as this lets people who evaluate the item being measured as average answer truthfully. This is called a Likert item. As a point of reference, Table 2.1 displays a few standard Likert tables to help you in your questionnaire building.

Table 2.1   Sample Likert Tables

For Measuring Agreement	For Measuring Frequency	For Measuring Likelihoods
• Strongly agree • Agree • Undecided • Disagree • Strongly disagree	• Always • Very frequently • Frequently • Occasionally • Rarely • Very rarely • Never	• Definitely • Very probably • Probably • Probably not • Definitely not
For Measuring Importance	*For Measuring Quality*	*For Measuring Value*
• Very important • Important • Moderately important • Slightly important • Not important	• Very good • Good • Acceptable • Poor • Very poor	• High • Moderate • Low • None

This scale can also be paired with a visual representation. For example, some survey platforms will allow you to ask a question and include a visual representation element as a grading scale (Table 2.2).

Remember that points on the scale should be labeled with clear, unambiguous words, and write questions so that both positive and negative items are scored "high" and "low" on a scale. Also, it's important to be consistent in building scales throughout the questionnaire: don't go from low too high in

Table 2.2   Example of a Closed-Ended Question Using a Visual Representation of a Grading Scale

	Very important				*Not important*
How important is price when purchasing products to help relieve the side effects of radiation and chemotherapy?					

one question, and then switch from high to low in the next one. Finally, building items that are symmetric (meaning they have the same number of positive and negative items) ensures a neutral data collection tool.

2.3.2 Open-Ended Questions

Open-ended questions are questions that require clarification from the participant. They require him to engage and justify his position, while closed-ended questions are questions where the participants only have a number of possible answers. Consider the following two examples relating to headache medication:

✓ Question #1: Enumerate all the different brands of headache medication you know. (open-ended)
✓ Question #2: Do you know the headache medication [XYZ brand]? (closed-ended)

In the first question, the researcher can generate a lot of information, such as

■ The consumer's knowledge of different brands
■ Which brand(s) comes to mind
■ Which one came to mind first
■ In which order they came to mind

■ Which one(s) the consumer does not have any awareness about

In the second question, the researcher is learning exclusively about the consumer's knowledge (or lack thereof) of the [XYZ] brand.

If there are a large number of people to question, the data from open-ended questions will need considerably more resources to be analyzed and codified, while data from closed-ended questions can be analyzed quite rapidly. Hence, budgetary concerns definitely weigh in when choosing between open- and closed-ended questions.

The other issue with open-ended questions is that they require more effort from the participant, giving him more "room to act" and answer falsely. Some participants take the opportunity given in an open-ended question to share displeasure with the product, the company, or even the survey, even if it has no relevance to the question. You can easily have 5% of responses in open-ended questions that are incoherent and have to reject them.

2.3.3 Projective Questions

Projective question techniques are used when the market researcher wants to get more insight into a participant's perception of a brand or product. Usually, participants are asked to project their thoughts onto other things, and are then asked to explain their answer. This probing into the participant's thought process can give more insight into the reasons they made these associations. Common projection techniques include

a. *Word association questions*: For this type of question, participants are given a list of words (or a series of pictures) and are asked to respond with what comes to mind. This gives the researcher insight into the consumer's vocabulary in relation to a brand or product. Using the results of multiple answers, researchers can uncover market trends.

b. *Sentence and story completion*: This involves asking a participant to complete a sentence or story. This helps to gain insight into the participants' predetermined attitudes and thoughts. It is mostly used when direct questioning is unlikely to provide an honest response (e.g., the participant might be too shy to answer truthfully). This type of question is useful in brand positioning and commercialization. For example:

■ People who use Brand XYZ are _____
■ I only buy product XYZ when _____

c. *Photo sort questions*: This technique involves giving participants a series of photos, then asking them to select which ones they associate with a product or brand. Afterwards, the participants are asked to explain why they chose these pictures to gain additional insight. It can very useful when identifying a product's perceived positioning, such as a medical device or laboratory product.

d. *Third-person technique*: This technique involves asking participants what they believe other people are thinking, feeling, and saying. This is useful when tackling subjects that are sensitive in nature, and you expect people might hide their true feelings on a topic. As such, topics in healthcare that are very sensitive (assisted suicide, abortion) are often best explored using the third-person technique.

There are other techniques that are used in market research but are less likely to be used in life science market research, such as

■ Cartoon drawing (presenting the participants with a cartoon and asking for interpretation)

- Personification (turning a brand into a person, and describing said person)
- Stereotyping (presenting a description of people and asking questions about the situation of interest)

2.3.4 Choice Modeling Questions

If you are trying to ascertain an individual's decision process, it can be a good idea to include choice modeling questions.

Choice modeling consists of asking a series of discreet questions where the participant makes relative choices (e.g., A vs. B, B vs. C, and A vs. C) in order for the researcher to infer the prioritization between A, B, and C.

To build a choice model, you have to properly identify the service or product to evaluate, and then select the attributes you are testing. For example, if you are creating a portable diagnostic device, you might be interested in learning whether speed, accuracy, price, robustness, or portability is the most important attribute. By asking a series of questions opposing one attribute to another, you can identify a pattern. It is possible to build more complex models, but for market research purposes, a simple model can often supply a wealth of information.

The advantage of choice modeling is that it forces the participants to make choices: asking if speed, accuracy, price, robustness, and portability are important will most likely leave you with inadequate data (they are all important after all), but having a series of questions that force the user into making a decision can more likely bring the consumer to reflect, and can generate a more definitive model. It can also be used to estimate the impact of pricing on attributes, and reduces the participants' ability to bias the research results.

The main disadvantage is that the data generated is ordinal, so it provides less information than ratio or interval data.

So you might learn which attribute is the most interesting, but won't necessarily know how much more important it is.

Overall, the main use of choice modeling is for predicting preferences and refining new product development, estimating the willingness of customers to pay for goods and services, estimating the impact of product characteristics on consumer potential purchases, and evaluating the importance of product attributes.

2.3.5 Question Banks

Question banks can be a great source of inspiration when building a questionnaire. Some sources for pre-existing questions are shown in Table 2.3.

Table 2.3   Online Websites Featuring Banks of Pre-existing Questions

Name (Website)	Comments
The General Social Survey (www.gss.norc.org)	Contains a core of demographic, behavioral, and attitudinal questions, plus topics of special interest
American National Election Studies (ANES) (www.electionstudies.org)	ANES produces high quality data from its own surveys on voting, public opinion, and political participation
Enterprise Surveys (www.enterprisesurveys.org)	Cover a broad range of business environment topics, including access to finance, corruption, competition, and performance measures
Association of American Medical Colleges (www.aamc.org)	Some useful question examples from surveys done in the last few years

2.3.6 *Mindful Surveys*

An innovative approach to designing questions for surveys is the Mindful Surveys Technique.* It integrates qualitative and quantitative questions into a two-step process to build an iterative questionnaire. This is especially useful for companies with a limited budget, or for researchers with limited knowledge of the topic that will be investigated.

As mentioned earlier, analyzing a large qualitative data set can be long and costly (especially if you have to do a double-bind evaluation by two independent researchers). A way to mitigate this is to use closed-ended answers, but you might not feel confident enough to enumerate all possible choices. In this situation, you might prefer to collect more information before developing a closed-ended question.

Mindful surveying is a two-step process to resolve this issue. The first step is to open the survey to participants, using open-ended questions, and collecting about 100 responses. Once enough completed surveys are received, the researcher temporarily closes the survey and analyzes the data. Using those responses, the researcher can identify patterns and trends to craft a closed-ended question. He can then modify the survey with his new question, open up the survey again, and collect the data in a more digestible format. If he could not identify enough trends, he could open up the survey again, collect another hundred entries, and analyze and try to identify trends again.

The researcher should remember when using this approach that it is important to set guidelines on the "stop-go sample" so that some groups are not underrepresented. For example, collecting 100 responses could be an end point, but if the researcher is surveying different age groups, it could be more relevant to stop the survey once he gets 50 answers for each age group.

---

* Henning, J. 2016. The Qual Sandwich Doubles Down on the Survey. Insight Associates. http://www.marketingresearch.org/article/qual-sandwich-doubles-down-survey (Accessed December 25, 2016).

2.3.7 Things to Watch For

Here are a few items that one should watch for when formulating questions:

a. *Double-barreled questions*: A double-barreled question is a question that is attempting to measure multiple items simultaneously. This makes it difficult for participants to answer, and leads to incorrect or unusable data. For example:

■ What is the quickest and most precise test to verify XYZ?

Are you interested in the quickest or the most precise test? What if one test is the quickest and another one more precise ... which one should the participant answer? Respondents who think one test is quickest and another more precise would be unable to answer, and if they do answer, the answer they would supply would be a compromise. Instead, the correct solution would be to split the question into two distinct questions, each having its own objective.

b. *Choosing the right words*: One of the biggest issues is the choice of words. Incorrect words can lead to bias, rendering data useless. Consider the following question:

■ Do you like it when your medication relieves pain quickly?

Very few people do not like quick pain relief, but the question as it is worded leads the person to answer in a very specific way. A better question would be

■ What are the characteristics you look for in pain medication?

c. *Overlapping interval answer sets*: Overlapping interval answer sets occur when categories are incorrectly designed, leading one or more answers to overlap. Consider the following question:

■ How many times do you see a doctor a year?

    i. *Never*
   ii. *1–3 times a year*
  iii. *3–10 times a year*
  iv. *10 times a year or more*
     See the issue? The categories overlap, so it would be impossible for someone who sees a doctor three times a year to answer this question. While seemingly innocuous, a researcher would be unable to use this data set.

  d. *Avoid technical terms, acronyms, and jargon*: Words should be easily understood by anyone taking the survey. But if you use a technical term or an acronym, a participant might be unable to answer your questions. Do not assume that participants have your level of technical knowledge, and instead use clear terminology. As such, rather than saying

■ Is arthralgia frequent at your clinic?
  Try

■ Is joint pain a frequent condition at your clinic?
  Another alternative would be to take the time and explain the acronym or complex term in your question, rather than assuming the reader already knows what you mean, but beware that many participants will just skip over the explanation and try to answer the question as best they can.

  e. *Avoid negative questions**: A negative question is one that is worded in such a way that it requires a "no" to respond affirmatively, and a "yes" to respond negatively. Questions that are phrased negatively rather than being phrased

---

* An interesting theory proposes to ask negative questions when a participant's privacy is a key concern. It argues that negative questions allow participants to keep the target data undisclosed by asking them, instead, to make a series of decisions with the information in mind. In this way, for example, the frequency of drug use can be calculated without respondents admitting to using any drugs. For more information, I suggest taking a look at "Surveys with Negative Questions for Sensitive Items" by Fernando Esponda and Victor M. Guerrero (2009).

positively are more difficult to answer and are more likely
to be answered incorrectly. As such, rather than saying

■ Do you believe the government does not do enough to
promote healthcare?
Rather try

■ Do you believe the government does enough to promote
healthcare?

f. *Thoughts on "prefer not to answer"*: Some researchers prefer to compel participants to answer every question and
do not include any "prefer not to answer" options in their
questionnaires. Others believe including a "prefer not to
answer" option increases response rates. Use your judg-
ment: if the question is very sensitive, include a "prefer
not to answer", which will ensure that the participants'
answer truthfully (rather than answering randomly to get
to the next question or dropping out of the survey). If
they prefer not to answer, at least the rest of the data you
collect will be useful.

g. *Integrate some consistency checks into your questions*:
Sometimes, users will answer your questions half-
heartedly, more preoccupied with the incentive than the
questions they are answering. Others might under-eval-
uate the commitment needed to answer all the questions, and start botching answers: this is especially true
if you are running a long web survey. To avoid collecting inconsistent data, integrate some consistency checks
into your questions to ensure your participants are
answering seriously, and the data you have collected is
useful.

This could include asking the same question twice
in the survey, and then comparing the responses
between both questions to ensure that the participants
answered the same way twice. Alternatively, you could
incorporate a simple "checking" question such as "If
you are reading this question, select 'Maybe' as the
response."

2.4 Incentives

Incentives are an important element to account for when budgeting for a market research activity. A properly structured incentive can increase participants' response rates by 5% to 20%. Incentives can take the form of money, information, services, and more. In an article published in *Health Service Research*, the authors found that the response rate of doctors to a survey varied significantly depending on the form of remuneration offered.*

In the study, doctors were invited to participate in a survey. Randomized groups were offered different incentives, all of which were worth U.S. $25. The incentives were immediate cash remuneration, immediate check remuneration, a promised check not requiring a social security number (SSN), and a promised check requiring a SSN. The immediate cash group had the highest response rate (34%), followed by the immediate check group (20%), the promised check without SSN group (10%), and the promised check with SSN group (8%). Globally, the study found that direct incentives generated more immediate responses than promised incentives.

The amount of the incentive also has a significant impact on the response rate. A recent study (2016) found that offering a modest compensation had little to no effect when surveying doctors. In that study, the response rate between those who were offered a book for participating in a survey versus those who were not offered a book showed little variance (11.6% vs. 10.8% response rate).† Another study found that offering a more significant gift ($50 redeemable gift card) was an

---

* James KM, Ziegenfuss JY, Tilburt JC, Harris AM, Beebe TJ. 2011. Getting physicians to respond: The impact of incentive type and timing on physician survey response rates. *Health Services Research*. doi:10.1111/j.1475-6773.2010.01181.x.
† Cook DA, Wittich CM, Daniels WL, West CP, Harris AM, Beebe TJ. 2016. Incentive and Reminder strategies to improve response rate for Internet-based physician surveys: A randomized experiment. Eysenbach G, ed. *Journal of Medical Internet Research*. doi:10.2196/jmir.6318.

effective way to motivate clinicians to participate: out of the 117 clinicians who participated in a survey, 63.5% of participants redeemed the gift card that they received, while the others did not redeem it. The authors believed that this reinforced the need for adequate compensation.*

There are a number of possible incentives you can offer participants such as:

a. *Direct remuneration*: A cash incentive can be offered to the participants. The Market Research Society† advises that any monetary incentive used should be "reasonable and proportionate," which means that incentives should be considered project by project based on the demographics of the expected respondents, how specialized the subject matter is, and how much the respondent will be inconvenienced by participating. The minimum you will pay for an in-depth interview will vary with the profession of the participant as well as the length of the interview: it could range anywhere from $50 for a non-specialized participant to $350 for a specialist doctor. In the case of a focus group, the cost per participant could range from $100 for a regular participant up to $300 for a specialized participant (e.g., a patient with a rare disease). Professionals could cost almost double in some situations, making this quite a costly endeavor.

b. *Vouchers*: A variation on direct remuneration, vouchers that are redeemable for specific goods are also well appreciated as they are perceived as having monetary value (a $50 voucher buys $50 of "real stuff"). The advantage for the researcher is that online vouchers are quicker

---

* Chen JS, Sprague BL, Klabunde CN, et al. 2016. Take the money and run? Redemption of a gift card incentive in a clinician survey. *BMC Medical Research Methodology*. doi:10.1186/s12874-016-0126-2.

† The Market Research Society (MRS) is a UK-based market research association for practitioners. It produces a series of guides on a number of topics interesting for market researchers. Please visit their website (www.mrs.org.uk) for more information.

and easier to purchase and distribute (this can all be done online), easier to track and manage (vs. checks or cash), and (sometimes) if bought in bulk, discounts are available.

c. *Non-monetary gifts*: Giving participants a non-monetary gift is especially advantageous in recruitment when there is a convergence of interests between the topic, the gift, and the profile of the participants. For example, discussing traveling trends with frequent travelers, and then giving them traveling gear as a thank-you for participating. The gift has to be carefully chosen to appeal to the participants (to make sure it isn't something the participants probably already have, and to make sure that it does incentivize the participants). Also, there could be some logistical concerns around managing the distribution of products (especially if you have to mail them), but there is the advantage of being able to purchase products in bulk.

d. *Free food*: This works best for people participating in a focus group while in a work capacity (i.e., employees) or in a community/patient group. You can buy a nicer meal than they would usually get for themselves, and it's convivial to "chat over lunch." In these cases, it is a nice touch to add a small participating gift (such as a small voucher) as a takeaway for participation.

e. *Charitable donations*: Some individuals with high revenue are insensible to monetary incentives. As such, offering the budget that you planned for incentives to a charity organization instead might incentivize them to participate. For example, a per diem of $100 might not be enough to encourage doctors to participate in a focus group, but donating the $2000 recruitment budget to a worthy cause and sharing this information during recruitment can sway and increase participation. It appeals to a participant's generous and philanthropic side. It is also possible to offer participants a choice in which charity the funds will be donated to. Once the donation is made, it is important to send a confirmation from the charity (or a contact at the charity itself) to participants so they know it has been done.

f. *Sharing the research results*: Some participants might be interested in the results of your findings, as the topic is directly related to their work. As such, it is possible to recruit participants by offering to share the high-level results of the research. In this case, you will have to prepare an executive summary of your findings specifically for those who request it.

g. *Bartering*: Also called the *Tepoztlan interview strategy**, this strategy consists of helping a group or a participant in a specific task in exchange for participating in the data collection effort. This could consist of assisting a non-profit trade association with a specific task in exchange for assistance in recruiting participants. This technique is especially useful when working inside a closed environment, such as a hospital, an association, or a community. The advantage is that while the researcher is rendering a service, his presence within the environment is much more understandable and enables him to be a participant observer.

Finally, some of the best practices related to incentives include

1. *Speedy delivery*: Incentives should be delivered between 1 and 2 days after the data has been collected. Participants shouldn't have to wait for weeks for their incentive to arrive, so don't wait until the end of the project to send incentives: it will be a lot simpler to manage if they are distributed as the interview is completed.

2. *Branded delivery tool*: The e-mail or physical mail that delivers the incentive should have the organizing company's information clearly identified so that participants

---

* This is named after a study that Oscar Lewis conducted in 1951, where he found that to establish a good rapport with the people he was studying in the Mexican village of Tepoztlan, it was necessary to help them.

can quickly identify the origin, and don't discount the
e-mail as spam when they don't recognize the sender.

3. *Properly crafted message*: The message that accompanies
the incentive should include a reminder of why they are
getting an incentive, and a thank-you note to make them
feel appreciated.

The benefits of respecting these best practices include mini-
mizing time doing customer service with participants looking for their incentive, less time spent resending lost (deleted)
incentives, and increased chances of the participants participating in future projects. If you are spending too much time
managing your incentive programs, look into some of these
tips or into automating some the incentive rewards you use.

2.5 Data Collection Methods

In this section we will be going over the basics of many market research data collection tools. These overviews are useful
to understand and select the instrument that you need for your
project. The methods we will be reviewing are in-depth interviews, focus groups, traditional surveys, online surveys, Delphi
groups, observation, and mystery shops.

2.5.1 In-Depth Interviews

In-depth interviews are interviews that are done one-to-one,
between the researcher and the participant. These interviews
consist of mostly open-ended questions. The objective is to
explore the topic in a semi-formal format, gathering qualitative
information.

The person interviewed could be a potential client, a
past client, a key opinion leader or anybody else with
a relevant relationship to the research topic. Current
employees are another good source of information as

they are in constant contact with clients and partners. As such, they can provide insight on current customer profiles, goods and services that are popular, satisfaction with current price levels, as well as experiences with competitors.

While in-depth interviews are more costly (in terms of time and money), they present a number of advantages over questionnaires and online surveys. For starters, interviews give more opportunities for the researcher to motivate the respondent to participate in a truthful manner, and to not abandon the interview halfway through. Also, interviews allow more flexibility in exploring topics as opportunities in the interview occur. The more exploratory the topic, the more useful the in-depth interview is, as it allows the researcher to change the order of questions, to prioritize some topics if time is short, or to go into more depth if the participant is discovered to be someone with a wealth of information on a specific topic. But the interviewer should be wary of using his interviewees' prerogative too often, since deviating from an existing script too many times can lead to a lack of standardization in data collection and issues in data compilation.

Also, interviews allow more control over the answer process. For example, it might be crucial for a participant to answer a series of questions before another, and an in-depth interview makes it easier for the interviewer to control the flow of responses. Finally, interviews allow the researcher an extra opportunity to evaluate the validity of the answers being given, by evaluating non-verbal cues, or by asking questions again later to validate consistency. This is especially useful when the topic is sensitive, and the researcher expects participants to exaggerate or undermine some facts.

Interviews allow researchers to gain additional insights on services or products being developed. Some topics that can be explored include

1. *Purchasing process*: How does the interviewee currently purchase the product? What are the criteria that he looks at the most? Which are the elements that are the most important in the purchasing process? Which things are not important? How important is price in the selection process?
2. *Current product*: What does the participant currently use? Why has he chosen to use this product instead of another? Has he tried other products? Why doesn't he use these products instead?
3. *Ear to the ground*: What has the participant heard about competitors? Has there been any movement or changes lately?
4. *Innovation*: What is lacking in current products? Is there something the product used to have and doesn't have anymore that irritates the participant? What innovations are coming in existing competing products? Anything that excites him?

On the importance of interviews ...

It is very valuable to talk with clinicians and nurses, who are directly working in the environments where the product is going to be used. It doesn't necessarily have to be formal interviews with recognized key opinion leaders at first: regular personnel provide priceless information as well. These discussions add a lot of value to the presentation, as the results give an early indication of market interest, as well as illustrating the gap the product aims to fill

Elizabeth Douville, Partner, AmorChem Financial Inc.

2.5.1.1 Keys to Performing a Good Interview

There are a number of actions a researcher can take to enhance the interview process. Here are a few useful suggestions.

- *Prepare your interview*: Conduct a quick due diligence check on the interview target prior to your interview, to identify potential specialization and fields of specific interest.
- *Record the interview*: When you make arrangements to record the interview, you will be able to focus on the answers the participants are providing, as well as asking questions and exploring topics with your interviewee, rather than spending time taking notes. If possible, use either a portable recorder, or if the interview is done by phone, use a service such as *Save Your Call* (www.saveyourcall.com). These services charge a minimal amount to record interviews, and you can listen to the interviews again later to transcribe, codify, and analyze them as needed.
  - *Remember to ask for consent*: Recording an interview without advising a participant is unethical. Furthermore, most people are generally fine about being recorded, and appreciate being asked beforehand.
- *Be clear on who and what the study is for, and what the objectives of the interview are*: It is important to share information on what the interview is for, why it is being done, and its purpose at the start of the interview. While the knowledge of who is sponsoring the study might polarize some participants, it is essential that the participant be advised upfront. Also, it is good practice to remind participants of the objective of the interview. Both of these kinds of information can diminish participant suspicions about how the information will be used.
- *Concentrate on the essentials*: Avoid long interviews. A good interview lasts as long as it takes, but when preparing your interview guide, plan accordingly. A good rule of thumb is that an interview with an unpaid participant should last 15–30 minutes at most, while a paid participant can expect to spend as much as an hour being interviewed.
- *Listen to your interviewee*: You are trying to collect data, so that means letting the participant share his

information. Be careful not to spoon-feed the answers that you are looking for. This leads to bad data collection, and ultimately, does not reflect real market conditions.

■ *Use silence to get more information*: Sometimes, getting the interviewee to share means not speaking for a few seconds: research in social interactions has shown that silence is as meaningful in verbal communications as rests are in a musical score. Inserting silences in an interview has a number of benefits: it slows the pace, it is conducive to a more thoughtful mood, it allows the respondent to control the direction of the next steps of the interview, it allows the researcher to note useful associations, and it reinforces the notion of interest on the part of the researcher as he is not perceived as rushing through the interview.

■ *Keep neutral*: This is especially hard when the subject of the interview is getting criticized or challenged (and you have a vested interested in the topic), but a good interview needs to be completed in a neutral fashion. As such, the participant should not be able to guess if you agree or disagree with his answer.

■ *Be encouraging*: This might seem to be in contrast with the advice to keep neutral, but the two are not mutually exclusive. Being encouraging means encouraging the speaker as he shares the information with phrases such as "that's interesting" or "can you share more about that?" Hence, the goal is to keep the interviewee engaged without influencing him by taking a position, or by sharing that position.

2.5.1.2 Different Types of Individual Interviews

There are different types of individual interviews, each with their own strengths and weaknesses. We will be touching briefly on face-to-face, phone, and online interviews.

a. *Face-to-face*: Face-to-face interviews are useful when dealing with complex topics. Also, when facing the person you are interviewing directly, you are guaranteed that the person is dedicating all of his attention to you. To minimize distractions, you can do the interview in a comfortable setting. Other advantages are that this allows the researcher to use visual stimuli (a mock-up of a potential ad or a range of potential packaging) and allows for longer interviews than phone interviews due to the individual proximity. Finally, the personal interaction allows for a more dynamic situation, keeping the participant motivated. The main disadvantage is the cost: it can be costly to travel to each interview (especially if they are geographically dispersed), or to bring every interviewee into your office, and it can be complex to compile a high number of interviews due to the complexity of the information gathered.

b. *Telephone*: Doing interviews by phone is both faster and more cost-effective than face-to-face interviews. This method is used in situations where multiple people will need to be contacted rapidly, or when you have a limited budget while still wanting the flexibility of an open-ended questionnaire. The main issue is that there are attention issues as the user can often multitask during your interview. Also, phone interviews are often missed by participants: as a rule of thumb, one or two out of each five interviews you book will most likely need to be rebooked due to a missing participant (especially if you are not offering an incentive). Topics handled by phone questionnaires have to be a little less complex, as you are unlikely to have the users' complete attention. Finally, it is difficult to pick up on non-verbal cues when doing a telephone interview, so you have to be particularly attentive to verbal cues.

c. *Online*: To mitigate the weaknesses of the first two approaches, it is possible to do interviews online using software such as Skype, Google Hangouts, or Viber. When doing interviews online with a camera, the researcher can make sure that he has the complete attention of the participant while also being cost-effective. Also, some technologies allow the researcher to record the interview for future playback and analysis: for example, some third-party providers have created applications that can be used to record Skype calls.* The downside of doing interviews is mostly technology limitations: quality in audio and video can vary (especially for some countries where the telecommunication infrastructure is relatively weak), so you can get intermittent errors that limit the information you get from visual cues, or lead you to miss some comments due to audio problems.

2.5.1.3 Motivating Interview Respondents

Some interviews will be more difficult than others. Some individuals will have changed their minds about participating; some might do it for the incentive, while others might be compelled (by a superior or colleague) into participating. Whatever the reason for the resistance, it could make the interview difficult and quite uncomfortable. Here are some tips that might help motivate the respondent and ensure that you get the information you are looking for.

■ *Increase the participants' interest*: A participant might have low interest since he fails to see the relevance of the answers he is sharing. Supplying context on why the interview is occurring, and mentioning the connection between the individual and the project can motivate the respondent. For example, reminding the participant that the objective of the project is to develop a technology

---

* A list of third-party applications recognized by Skype is available at support. skype.com/en/faq/fa12395/how-can-i-record-my-skype-calls (Last visited January 1, 2017).

that will simplify the participant's daily work could help motivate him.

■ *Recognizing the participant's expertise*: Reminding the participant that his expertise was the reason he was selected for the interview can enhance his need for recognition. This can be done not only at the start of the interview, but throughout the interview as well.

■ *Reducing threats to the ego*: If you feel the participant's ego was bruised by a previous question, it is a good technique to include a face-saving preface to attune the impact. For example, you could start a new series of questions with a statement like: "As you know, topic ABC is not a problem many researchers are familiar with. But I believe you have some insight on it, so I'm looking forward to hearing you thoughts and experiences on it."

■ *Reducing the issues related to perceived authority*: There may be situations where the participant identifies the interviewer with a role or position that influences the interview process. For example, the interviewer could be a key opinion leader, or a former teacher collecting data from past students. Clarifying the interviewers' role in the interview process throughout the interview can enhance the interview flow.

2.5.1.4 Interviews in Life Sciences

In-depth interviews in life sciences are a popular way to get the information you need, especially when dealing with topics that are sensitive in nature. It is a great way to speak to people confidentially and get their views on healthcare topics, such as their personal health and that of loved ones, their use of pharmaceuticals, and so on.

Interviews are also useful to discuss topics for which an "inch-deep, mile-wide" approach is needed. Sometimes, a market researcher will have little information on the topic he is researching, and asking a few general questions will enable him to quickly identify opportunities.

Interviewing doctors and medical personnel can be especially challenging as these individuals are constantly solicited for their time. As such, you might have to set aside an important per diem to interview them, ranging anywhere from $150 to $300 per interview (and even more for key opinion leaders). Here are some things to keep in mind:

■ Identifying doctors to interview can be challenging. One way to save time is to purchase a list from a list provider. You can customize your purchase to your needs (by specialty or geographic region). Be sure to double-check how often the data is updated as well as what contact information they will be supplying (e-mail, physical address, phone, fax number, etc.). Also, verify if the doctors have opted in to be contacted by third parties, or if the information was gathered from the web at large (i.e., scraping).

■ If you want to build your own list, professional associations, licensing boards, or lists of attendees at a conference are good starting points. Reviewing scientific publications through a database such as Google Scholar, PubMed, and Europe PMC can also be useful to identify key opinion leaders and topic experts, as well as obtaining their contact information.

■ The best way to reach doctors is by e-mail. Phones calls often end up in voice mail or are intercepted by gatekeepers (such as administrative assistants).

A typical e-mail might look like the following:

I am conducting a study on behalf of Bio Biotech on the technologies currently used to diagnose XYZ. The results of the study will be used to help our client *develop a new diagnostic device to better meet your and your patients' needs. We value your input and would like to arrange a 20-minute telephone phone call to discuss the topic with you. Please note that all results will be anonymized, and only consolidated results will be shared with the client. In recognition of the value of your time, we are offering an honorarium of $$$.*

Another way to recruit specialized participants is to use one of the many online platforms that exist where experts are available for consults and single interviews. The platforms, such as Zintro (www.zintro.com), enable you to easily post projects and contact subject-matter experts directly, acting as an escrow service. You can find scientists, doctors, and more on these platforms, and experts can either respond to your enquiry or refer it to other experts in their network.

Finally, don't get discouraged if you get low traction for your solicitation: remember that doctors are a heavily solicited population, and that it will take a large recruitment effort to obtain your target sample.

2.5.2 Focus Groups

A focus group is a small group of individuals brought together to discuss a specific topic. The added value of a focus group (vs. individual interviews) is that the interaction between individuals generates a wealth of information for the researcher. As such, focus groups are useful to gain a better understanding of what people are thinking, and also why they are thinking it. It is also interesting as many participants will generate more information in a group setting where they feel safe and they do not feel that they are the sole focus of the interview. Finally, the information gathered in a focus group can be very useful to design a quantitative questionnaire, or in interpreting the information gathered from a quantitative research project.

The main issue with focus groups is similar to issues you find in peer groups. As such, social pressure (the desire to conform to the group and not disagree), an individual's domination of a group, or the halo effect generated by key opinion leaders can all impact the quality of the interactions and the information generated. It is the role of the moderator to step in and re-balance the focus group and ensure that it does not become biased. The other issue to remember is that while focus groups can be used to evaluate a group's feelings or views on a topic, it cannot be used as a final decision tool. It is exploratory in nature, not statistically valid, and the information you gather, while invaluable in interpreting existing data or setting up more research, cannot be an endgame in itself.

Ideally, a focus group will have 8 to 10 participants, as the group needs to be sufficiently large to generate dynamic conversations, but not too large as to leave some participants out or to become difficult for the moderator to manage (as parallel discussions start to be generated). Smaller groups (five to six participants) are also possible, but create a different group dynamic, and as such are more useful when focusing on a specific topic, or if they include a product-testing component.

About 45–75 minutes is the ideal time frame for a focus group: too short, and you will not generate any deep insight, while too long, you risk participant fatigue where they quickly agree in the hopes that the focus group will end. Three to four focus groups are usually necessary to fully explore a topic: after that, you may find the same information being repeated.

In life sciences, focus groups are especially useful for interacting with end users (doctors, nurses, laboratory technicians) as well as getting feedback from patients. As such, they can generate valuable information related to marketing, branding,

competitors, and product issues. Nonetheless, some researchers have found that focus groups are not an ideal environment
for eliciting emotional information from physicians, because
physicians' self-perceptions and the image they want to project to others are typically those of a rational (not emotional)
decision maker.* Also, it can be especially challenging to reach
and recruit specific participants.

2.5.2.1 Guidelines to Preparing a Focus Group

A focus group discussion centers on a predetermined set
number of questions (anywhere from 8 to 10 questions is a
good number of questions). It is important not to overload the
discussion guide with too many questions as the moderator
might not have time to tackle every topic or might feel rushed
to touch on every topic, missing some key insights.

When writing questions, keep them open-ended: participants should not be able to answer them with a simple "yes
or no." Also, participants usually do not have a visual copy of
the question they are discussing, so try to keep them simply
phrased. If necessary, ask follow-on questions to tackle complex topics. A good methodology to use when building your
discussion guide is to have three groups of questions:

1. **Engagement questions**: Simple questions to warm up
   the group, so they get to know each other and the topic
   of discussion, making them comfortable
2. **Exploration questions**: Main questions for the focus
   group on the topic the researcher is researching
3. **Exit question**: The last question to see if the participants
   want to add anything relevant to the topic before ending
   the focus group

---

* Kelly D, Rupert E. 2009. Professional emotions and persuasion: Tapping non-
rational drivers in health-care market research. *Journal of Medical Marketing:
Device, Diagnostic and Pharmaceutical Marketing,* 39: 3–9.

Table 2.4 provides a sample discussion guide for a focus group on store-brand headache medication.

Table 2.4   Example of a Focus Group Discussion Guide Exploring Patient Engagement with Store-Brand Headache Medication

Engagement Questions 1. What is your favorite brand of headache medication? 2. Which store do you usually buy headache medication from?
Exploration Questions 1. How do you choose which medication you will buy at the store? 2. Has anybody here ever tried store-brand headache medication? What was your experience with it? 3. What are your thoughts on store-brand headache medication? 4. How safe are store-brand medications compared with brand medications? 5. What are the pros and cons of store-brand medication?
Exit Question 1. Is there anything else you would like to say about headache medication?

2.5.2.2 Recruiting for a Focus Group

The participants' profiles as well as inclusion/exclusion criteria should be decided before the recruitment process begins. This information will be useful to screen and select participants. Examples of criteria include the sex of the participant, age, user versus non-user, or any other focus you have defined earlier in your market research objectives. There are many ways to recruit participants for a focus group: some examples are included here for convenience.

a. *Nomination*: Participating key opinion leaders can nominate other applicants they believe would make good contributors. Nominated individuals are interesting potential panel members as they are expected to already have a

good knowledge of the topic at hand, and are more likely
to want to participate as they have been referred by a
noted expert.

b. *Random selection*: If working with an especially large
pool of willing participants, the researcher can select
at random participants out of a pre-existing pool (e.g.,
employees in a company) until the right number of participants is attained.

c. *Members of a group*: In some cases, an existing group of
participants can be a great recruitment pool from which
to invite participants. For example, a non-profit association or a chamber of commerce can both provide interesting pools of participants.

d. *Volunteers*: When doing a broad research effort, participants can be recruited at large through traditional means
(flyers, newspaper ads) as well as electronic means, such
as Craigslist, online job boards, and specialized discussion
groups.

When forming your groups, it is good practice to over-allo-
cate and include one or two more participants than needed: it
is not uncommon to have 20% or more of a group not show
up (even if they are incentivized), so overbooking slightly mitigates this.

Also, do not forget the multiple costs associated with running a focus group, such as recruitment costs (anywhere
between $250 and $750 per panel if you are dealing with a
specialized third party), recruitment fees and incentive fees,
facility costs (room, food, recording), direct moderator costs
(which can range from $750 to $1500 per session for skilled
moderators), and indirect costs (transcription, analysis, and
final report). Complete focus groups can be a very costly
endeavor, especially in a start-up situation. The two variants
discussed later (online focus groups and triads) are interesting
alternatives if you are unable to shoulder the full costs associated with a traditional focus group.

2.5.2.3 Running the Focus Group

Ideally, a focus group is moderated by two persons: the head moderator and an assistant moderator:

■ The *head moderator's* responsibility is to facilitate the discussion and to directly intervene with the group. He listens and ensures that all participants participate. He should have solid knowledge of the topic to be able to ask follow-on questions. Skilled moderators will be able to paraphrase long and complex comments, and build on these to move the conversation forward. Finally, as the moderator is in a position of authority (he is moderating and handling traffic in the conversation), he must stay neutral at all times and refrain from agreeing or disagreeing with the comments as they are shared.
■ The *assistant moderator* takes notes and handles the recording (if the focus group is being recorded). He listens, notes, and observes any subtle events that might occur. It is preferable that he lets the moderator talk and handle the conversation, but must be ready to add his input if needed.

For analysis purposes, it is useful to collect participant's demographic information. A short questionnaire that requires a few minutes is all that is needed. The assistant moderator has to make sure that participants have answered the questionnaire before coming in to the focus group session, or has them do it on-site before the focus group begins (Table 2.5).

It is also necessary to have the participants fill out a consent form. The consent form informs participants that they have a responsibility to keep all discussions private as well as an agreement not to discuss any information discussed during the focus group outside the focus group setting. It can also include an authorization request to quote participants (anonymously or directly).

Table 2.5   Example of a Demographic Form for a Focus Group with Physicians

Physician Focus Group Participant Demographics	
Date:	
Time:	
Location:	
Your gender	Your age
• Male • Female	• 30–40 • 41–50 • 51–60 • Over 60
Type of practice • Public • Private • Other: _____	What is your specialty? • Cardiologist • General physician • Other: _____
How long have you been practicing? • Less than 5 years • 5–10 years • Over 10 years	How many CME credits did you complete last year? • Less than 50 credits • 51–100 credits • Over 100 credits

Once consent forms and demographic sheets are collected and reviewed, the focus group starts. The moderator welcomes participants, reminds the participants of the focus group objectives, and sets some ground rules. These ground rules could include:

■ I would like everyone to participate: Please do not interrupt another participant when they are sharing their opinion.

■ There is no right or wrong answer: Everybody is sharing valuable insights and information, so no judging someone's opinion.

- But you are free to disagree! If you agree or disagree,
  share it with us, we want to know! Tell us why, in a
  non-judgmental way.
■ This meeting is confidential. What you share with us is
  confidential, and when we share it with others, it will be
  done in an anonymous manner.
■ We will be taping the conversation: We want to be sure
  we don't miss anything. But don't worry, there are no
  names in the final report, and everything will remain
  anonymous.

Immediately after all participants leave, the head moderator and assistant moderator should take a moment to debrief
while the recorder is still running, while their reaction is still
fresh. This debriefing will be an invaluable part of the data
collected during the focus group.

2.5.2.4 Variation #1: Online Focus Groups

Online focus groups are a variation of traditional in-
person focus groups. As their name implies, online focus
group participants participate online using a webcam and
microphone.

The advantages are numerous. First, there are lower costs
associated with online focus groups: some estimates are that
online focus groups cost about half that of a traditional focus
group. Also, online focus group recruitment is faster, and
the time commitment is less intense. This is especially useful
with professionals (such as doctors) who have limited time to
participate in focus groups and are difficult to reach. Similarly,
individuals with specific characteristics, such as rare diseases,
may not be in close geographic proximity to each other.
Online focus groups make it easier to bring these individuals
together, and make it easier to reach a representative sample
size. Moreover, people who are shyer are more willing to talk
in an online setting as they feel less directly judged by others

(not being in the same room and not being in direct contact). Finally, observers, clients, and the assistant moderator can observe on a separate feed, and give comments to the moderator discreetly. Note that if the client is observing in a direct feed, notice should be given to the participants.

Nonetheless, there are a number of issues with online focus groups. For starters, as participants are not in proximity to one another, there is a limited opportunity for direct interaction. As such, a lot of the direct interaction (which is sometimes the richness of a focus group) is lost in an online setting. Some who are more critical of online focus groups will go as far as claiming that the interaction in an online focus group is insignificant. Also, populations who are technologically challenged may be unable to participate effectively in online environments. So, if you are working on getting feedback from an older population, online focus groups might not be an effective solution. Furthermore, technology limitations (poor webcam or poor microphone quality) can all have an impact on the quality of the focus group session, and it is a lot more difficult to ascertain a person's identity online than it is in person. Finally, it is quite difficult to act on non-verbal cues in an online setting.

To ensure best results when doing an online focus group, a number of guidelines need to be respected.

1. Groups must typically contain between six and eight individuals, but the researcher might need to over-recruit one or two participants to account for no-shows (as these are more common than in traditional focus groups). Also, during the recruitment process, the researcher should ensure that the participants have the minimum technology requirements and knowledge to be able to participate.
2. During scheduling, the researcher should account for time zone differences between participants when choosing the time slot.

3. It is important to test the equipment and software before the actual online session. For testing purposes, the researcher can recruit a few of the scheduled participants to participate in a testing session to ensure that the moderator, clients, and other stakeholders have an adequate handle on the technology.
4. Finally, the researcher should make sure that he sufficiently communicates with participants. This includes reminder e-mails and detailed explanations on how to log in to the system.

During the presentation, the moderator can use PowerPoint slides to share content, and he has to be sure to keep an eye on the participants in attendance. It is his responsibility (or the assistant moderator, if there is one) to make sure that participants are attentive and that they are not multitasking (checking e-mail, playing games). Some software will let the moderator know if the participants' window is active, making surveillance easier.

2.5.2.5 Variation #2: Triads

Triad focus groups consist of three participants, and are used in very specific situations.* As there are fewer participants, these can be shorter (getting everything you need done in under an hour), allowing for more groups to be seen in the same time period. Also, the presence of fewer participants allows the researcher to do more in-depth research and probing, while limiting groupthink (participants retain more of their individual position, and outliers are less influenced by the general agreement of a large group). Finally, triad focus groups allow more dedicated product testing and observational usage,

---

* Some moderators encourage the use of dyad focus groups (two individuals) instead of triads. These groups can focus even more on product testing and feedback, simulating a small work group in an organization. The downside is that dyads offer a limited source of information in terms of richness of context.

as each participant can have more time individually testing the product, rather than having a few participants test the product and the other participants observing it.

The main disadvantages are that since the groups are smaller, more of them need to be conducted to obtain enough information. Triads should not be used as a cost-saving measure, but rather as a different methodology to obtain the same information. As a rule of thumb, you should hold enough triad focus groups to equal one focus group (three triad focus groups for each regular focus group). Also, triad focus groups can be more "awkward" to manage, since there are fewer individuals, putting more pressure on the moderator's skills to keep the discussion going forward.

2.5.3 Traditional Surveys

Traditional surveys are a cost-efficient tool used to collect a large amount of quantitative data from a target audience. Using a questionnaire, the researcher develops questions that tackle the topic as needed: most questions will be closed-ended to allow the participant to complete the survey rapidly (and to allow for quick data compilation), but there is the possibility of having some open-ended questions as well. The two main advantages of traditional surveys are that they are economical and that they can be done in a manner to protect the identity of respondents.

Questionnaires can be prepared and distributed in a much more economical way than one-to-one interviews. For example, if a researcher is administering a survey by mail, or soliciting a large group of individuals at a conference, lesser costs are distributed over a larger sample of participants. Secondly, questionnaires can be sent to potential participants with a pre-addressed envelope, which participants can send back to the market researcher without any identifying features.

There are several different techniques that can be used to distribute and collect information with questionnaires. We will touch briefly on those used less often in life science, and we will discuss in depth the most useful one, online surveys, in the next section.

2.5.3.1 Direct Mail Surveys

Direct mail surveys are handed out physically and directly to participants (either by handing them out in person or by sending through the mail). They have long been the technique of choice since they are less expensive, have a wide geographic reach, and there is no risk of interviewer bias. A recent 2016 survey by Accelerant Research found that 44% of respondents thought direct mail survey was an acceptable way of being contacted for a survey.* Nonetheless, their current effectiveness is questionable (direct mail surveys usually have a 2–4% response rate), there is a lengthy response cycle (between the survey being first sent out and getting the response back to the market researcher), and you need to engage in constant reminders (by phone or e-mail) to increase participation.

They have been largely replaced by online surveys but could be useful if addressing an older population that is less technology savvy. To increase participation, a Media Logic study found that envelopes which grabbed attention, seemed friendlier, and stood out more were more likely to be answered by seniors in a healthcare setting. Letters that look easier to read, are quicker and easier to understand, and have a better layout were more appreciated.† Another way to increase participation is to include a pre-paid envelope

---

* McDowell B. 2016. How to Succeed in Survey Invitations (According to Respondents). Insights Association. http://www.insightsassociation.org/article/how-succeed-survey-invitations-according-respondents (Accessed January 19, 2017)

† Media Logic. 2016. Medicare Marketing Insights: What Direct Mail Designs Do Seniors prefer? June 28th 2016, http://www.medialogic.com/uncategorized/blog/medicare-marketing-insights-direct-mail-designs-seniors-prefer/ (Accessed December 11, 2016).

as well as personalizing the letter that accompanies the questionnaire.

2.5.3.2 Telephone

Telephone surveys are a cost-effective way of collecting data, especially if you are trying to quickly reach a population at large. They usually exhibit higher response rates than online surveys, enable fast collection of data, can be used to tackle complex topics, enable specific targeting of participants, and usually have more positive responses. Telephone surveys' average response rates can vary from around 8% to 12%.

Also, most phone interviews are recorded, which makes it easier to validate interview quality and to transcribe and analyze data. Nonetheless, there are fewer and fewer landlines, making populations harder to reach. Furthermore, participants are more likely to drop out halfway through or give fake responses (especially if the phone survey was longer than initially announced). Finally, participants are less likely to answer if they do not have a relationship with the caller (i.e., they do not recognize the caller ID) and timing is everything when doing a phone survey: calling participants at an inappropriate time will most likely result in the caller refusing to participate.

Companies and organizations that have pre-screened lists (members, clients, and opted-in individuals) are more likely to have successful telephone survey campaigns than those that make cold calls. Another way to increase participation rates is to prepare lists of questions that are simple to answer and to keep the survey short. Remember that if you are talking to people representing organizations, make a note of which individual you interviewed.

2.5.3.3 In-Person/Street Surveys

In-person surveys bear a resemblance to in-depth interviews, but focus on quantitative data. They consist of approaching people in the street or in a natural setting and asking them for a moment so you can ask them a few questions. This is the main strength of in-person surveys: they can be conducted anywhere. For example, if you want to open a healthcare facility, or sell a product locally, it could be a good way to validate market demand by visiting the chosen location and surveying people around the location to get a sense of the "walk-in" traffic potential (number of potential clients who will enter a facility as they pass by it). This method is also useful to quickly collect data on a generic topic, and allows the use of follow-up questions, adjusting the questions in a fluid manner. It also permits the use of simulated materials and mock-ups as the participants are facing the interviewer.

However, it can be difficult and frustrating to recruit participants. Some things you can do to attenuate this include carrying an official ID, surveying the site in advance, and keeping the interview under 10 minutes. A small (visible) token of appreciation can also enhance participation. Another issue is that in-person interviews can take a lot longer to complete than a web survey. It is also more costly, especially if you include data compilation and the interviews you will have to reject due to invalidity concerns. Nonetheless, using an in-person survey allows some innovative practices (starting the survey in person and asking participants to complete it online at a later time or the use of mock-ups) and can be useful in very specific situations.

2.5.4 Online Surveys

The rapid pace and development of technology have created new opportunities for collecting data. As such, the use of online surveys has grown immensely in popularity. They are cost-effective, simple to use, and if done properly, can reach a wide range of populations and allow the participants to complete the survey on their own time with little effort. Also, they allow for the use of different media (such as sounds and videos), making the experience richer for the participant, and enabling more complex data gathering for the researcher. Most online surveys today are done using a web-based survey tool, although the popularity of surveys on a mobile platform is rising: this has the advantage of allowing for participation tracking (enabling you to directly target non-respondents and send them reminders), while allowing for direct data entry, making them much more efficient.

There are some limits to using an online survey for data collection. For starters, they are much more effective when used with a closed population (employees, members, clients) as the response rate will be higher. Also, online surveys are focused on populations that have access to web-based technologies: hence, Internet surveys will usually skew toward populations that are younger than 65 years old, college educated, and have higher than average household revenues. If you are not targeting one of these groups, another data collection tool might be more appropriate. Finally, with the rise in popularity of online surveys, there is a noticeable trend toward over-solicitation of users, leading to user fatigue: proper incentives (discussed in Section 2.4) are becoming a key factor in increasing user participation.

There are considerable challenges in recruiting participants for a life science survey, particularly clinicians. A study done in 2015 found that an online web survey targeting clinicians got a 35% participation rate, with deep variances across specialties: ranging from 46.6% (neurology/neurosurgery), 42.9% (internal medicine), 29.6% (general surgery), 29.2% (pediatrics) to 27.1% (psychiatry). Lack of time and survey burden were the most common reasons for not participating.*

Another study found that general practitioner survey rates could be increased with incentives (larger and upfront, if

---

* Cunningham CT, Quan H, Hemmelgarn B, et al. 2015. Exploring physician specialist response rates to web-based surveys. *BMC Medical Research Methodology.* doi:10.1186/s12874-015-0016-z.

possible), peers pre-contacting targets by phone, personalized packages, and sending the survey on a Friday.*

2.5.4.1 Tips for Designing Your Web Survey and Increasing Your Response Rates

There are several things to remember when building a web survey to increase your response rate.

1. *Keep the survey short and simple*: This helps to reduce user attrition. You might be able to utilize a longer survey if you are providing an incentive, but keep in mind that the size of the incentive is directly related to the length of the survey.
2. *Work on your formatting*: When designing your survey, make sure that it is visually appealing to participants and easy to read: the easier it is for the participant to read and navigate, the less attrition you will get throughout the survey.
3. *Be straightforward about the time to answer*: Announce upfront the length of the survey and, if possible, use a progress bar on top of the survey to keep participants engaged.
4. *Optimize your survey for mobile devices*: Use a survey platform that will optimize your survey for mobile platforms: more and more people use mobile devices to do mundane tasks while waiting in line or commuting, for example, and if your survey does not properly display on a mobile device, they might simply drop out and move on.
5. *Include an end of survey message*: The end of survey message should include a short thank-you, a reminder of the objective of the survey, and your coordinates or a link to a dedicated e-mail (such as survey@yourcompanyname.

---

* Pit SW, Vo T, Pyakurel S. 2014. The effectiveness of recruitment strategies on general practitioner's survey response rates—A systematic review. *BMC Medical Research Methodology*. doi:10.1186/1471-2288-14-76.

com) so participants who have concerns or questions can easily contact you.

6. *Send reminder mails*: Send the first reminder e-mail a week after the initial invitation and a final reminder the week after that. You should expect diminishing returns of about half each time you send out a reminder. So, if the first e-mail got you a 10% response rate, the first reminder should get you another 5% of respondents, and the last e-mail another 2%–3%.

7. *Plan the first survey invitation*: When sending out web surveys, target Mondays for B2B web surveys, as this will increase your response rate. If you are sending to end users, Wednesdays and Fridays are preferable to get the best response rates.

Finally, make sure you have informed consent and comply with anti-spam regulations. Informed consent is achieved by clearly explaining the survey objectives at the start of the survey, providing the identity of the sponsor of the survey, explaining how the data will be used as well as reiterating the confidentiality agreement surrounding participation in the survey. You can include a "do not agree" box at the start of the survey to discard participants who do not consent to the survey conditions. As for anti-spam regulations, these vary from one jurisdiction to the next, and only apply if you are surveying a population that has not opted in.

2.5.4.2 Online Survey Tools

There are many web survey tools available online that you can use to build, distribute, and collect data. It would be impossible to list them all here, so we are sharing a few of the most popular ones.

a. Survey Monkey (surveymonkey.com): Survey Monkey is probably the best known of all online survey tools. It is easy to use with both free and paid options and with its ready to use tools, a survey can be quickly implemented and made available online. Also, Survey Monkey offers services to recruit participants for your survey directly, respecting the demographic and characteristics you hand-picked for your participants. The downside is that some options cannot be found intuitively, so a user not familiar with Survey Monkey will have to poke around until they find the needed options.

b. Google Surveys (www.google.com/surveys): Google Surveys is an increasingly popular tool to build online surveys. There are two ways to create a form in Google Docs. The first way is to create a new form from Google Drive while the second way is to create the form from Google Sheets, which will link the spreadsheet to the form and load all the data into the sheet for later analysis. Simple and free, Google Surveys also has the advantage of having the data pre-formatted for the analysis phase.

c. Zoho Survey (https://www.zoho.com/survey/): Zoho Survey is a very simple to use online survey tool, with a quick setup and simple deployment. It is useful for doing quick surveys without a large budget, but it does lack some of the more sophisticated features of the other survey tools.

d. SurveyGizmo (www.surveygizmo.com): SurveyGizmo is a more advanced and powerful survey tool. The free version gives access to the most basic options, and the paid version gives access to more powerful tools, such as styling options, a question library, and e-mail campaigns. Of course, the more complex the tool, the steeper the learning curve.

e. Checkbox survey (www.checkbox.com): Checkbox survey provides advanced tools and customization. As such, it lets the researcher customize almost everything around his survey, while keeping costs reasonable. As with more complex tools, the researcher will have to invest more time learning for this tool to become functional compared with the more simple tools out there.

f. Pollfish (www.pollfish.com): Pollfish is a simple to use web survey tool that is perfect for beginner researchers. Once the researcher has posted his survey, someone from Pollfish revises the survey, offering helpful advice in optimizing questions and programming survey logic. The survey platform also recruits participants for your survey, enabling you to have your data quite rapidly.

2.5.4.3 Mobile Online Surveys

Mobile online surveys use mobile technologies (mobile phones, tablets, and PDAs) to collect data. The growth in popularity of these devices has created new opportunities to reach participants and collect data rapidly: Pew Global reports that an overwhelming majority of the population in almost every nation they surveyed owns some form of mobile device, even if it is not considered "a smartphone."*

Mobile online surveys present multiple advantages. They can be used to reach populations that are increasingly hard to reach with traditional market research methods (i.e., younger audiences). Also, they usually have higher response rates as mobile surveys are convenient to answer. They also have lower overall costs as it takes less effort to reach the minimum sample size, and mobile surveys can use some of the device's features (e.g., the global positioning system, the camera, or the microphone). Finally, as participants often have their mobile device on them, they respond quickly, leading to immediate feedback and even enabling event-based surveys.

---

* Pew Research Center. 2016. Smartphone Ownership and Internet Usage Continues to Climb in Emerging Economies. http://www.pewglobal.org/2016/02/22/smartphone-ownership-and-internet-usage-continues-to-climb-in-emerging-economies/ (Accessed December 11, 2016).

Some of the downsides of mobile marketing research are reduced response rates depending on the group targeted (some demographic groups are underrepresented on mobile devices), the questionnaire has to be shorter and simpler (making it difficult to collect in-depth data), and some practitioners have found that higher incentives are necessary to gain traction. Also, some people find receiving online survey invitations particularly intrusive (akin to getting a phone call during supper). The overuse of mobile online surveys could lead to members or customers opting out of getting contacted in future communications.

There are a variety of ways to collect participants for the survey, ranging from push methods (where the survey is sent out to participants independently of location and actions) to pull methods (the survey is sent to participants in a specific location, or those purchasing a specific product).

2.5.4.4 Getting Participants

Once your survey is built, the next step is to get participants to answer your survey. If you are surveying existing clients or interested parties, you can invite the participants directly to participate in your survey. If you need to survey people at large, there are a number of solutions that are available to you.

The first solution is to purchase a list or database of prequalified e-mails. A number of associations and private organizations sell lists of prequalified participants and databases that can be used for survey purposes. The price of these lists can vary a lot, from a couple hundred dollars to thousands of dollars. You have to remember that purchasing a list does not automatically mean that you have the authorization to contact the participants, so it is important to verify whether the database consists of participants that have opted in to receive communications from third parties.

Furthermore, if you are sending out your survey to a population at large without any consent of solicitation, you may run afoul of anti-spam regulations. Make sure you comply with existing regulations by adhering to local requirements (e.g., including a correct header and adding an unsubscribe link). If you are contacting individuals with whom you have an existing relationship (customers, members, employees), they are usually excluded from such regulations. Double-check your national guidelines to be on the safe side.

The other alternative is to use a third party to recruit participants. For example, Survey Monkey offers this service, and each participant will cost you $1–$7, depending on the criteria of the participants you need. Another alternative is Pollfish, which uses proprietary and third-party mobile phone applications to recruit and validate participants. They currently have a database of 370 million users throughout the world. In both cases, expect anywhere between 24 hours and 2 weeks to reach your target sample size. As you only pay per completed survey, not per respondent (e.g., you don't pay for incomplete surveys or those who did not pass the screening qualification), these recruitment services are quite affordable for start-ups trying to get a better perspective on their market.

As an example, in a recent project, a client needed to better understand the purchase process of skincare products for patients undergoing radiotherapy. The secondary information available on the topic was quite limited, so we engaged in a web survey online using a third-party recruitment service. The simple web survey was online in 24 hours, and within 1 day, we had a sample of 150 people sharing insights on their purchasing process, favored brands, and pricing sensibility for less than $300. This type of personalized market research would have been impossible just a few years ago, and speaks volumes on how emerging technologies are shaping the way we do market research.

2.5.5 Delphi Method

The Delphi method is an interesting methodology that is used to get information from experts and key opinion leaders on future trends. It is suitable when the researcher is faced with limited secondary information or a very complex topic.

The methodology is simple. A group of experts are individually asked a series of open-ended questions on a specific issue (usually by e-mail, although this was traditionally done by mail). The moderator then compiles the answers he has obtained, identifying the majority position of the group, listing the arguments for and against the positions. He then shares the majority position with participants as well as generating a new round of questions, which he sends out to the participants for additional feedback. As such, the participants are confronted indirectly, and face the opinions of other experts in an anonymous setting. This process is usually repeated until a consensus forms around the topic being discussed, but some researchers suggest that a maximum of four iterations should be done before concluding a Delphi group. Also, divergent opinions should be included in the report with their supporting arguments.

Hence, a Delphi group involves the systematic questioning of renowned experts to explore a breakthrough topic. This methodology is quite useful for forecasting trends, and for getting experts who are difficult to reach individually (as responses are completed at the convenience of the participant) while conferring anonymity on the participants. Typically, participants are not revealed to each other, even at the end of the project. The systemization of data collection combined with the anonymity given to participants minimizes the halo effect that some key opinion leaders can have on other participants, and enables participants to re-adjust their opinions more freely throughout the process. It is also relatively inexpensive.

However, the Delphi method results are tied to the competency of the moderator managing the group. An inexperienced moderator might lead the group to a false consensus, or lose participants as the group progresses. Also, there is a tendency to eliminate more radical options and focus on more mainstream compromises. Finally, the method is more time-consuming and lengthy than a traditional focus group.

In this method, the bulk of the work is done by the moderator. He is responsible for generating the initial questionnaire, following up with participants, compiling the information as it is generated, and animating the panel. He is also responsible for coaching participants and training them if they are not familiar with this methodology. His skills in written communication will have a direct impact on the success of the panel.

The researcher should be careful when recruiting participants and he should make sure they have been informed that the process is time-consuming. He should account for participant attrition, and plan accordingly by initially over-recruiting. While some researchers advocate for the recruitment of up to 30 participants initially, too many participants lead to complexity issues in data compilation and coordination: up to 15 participants is a good starting point, which takes into account participant attrition.

One thing to note is that the Delphi method is recommended for use in a healthcare setting as a reliable means of determining a consensus for a defined clinical problem, but might have limited practicality in market research situations looking to generate some market-oriented data (such as market size and growth).

2.5.6 Observation

Collecting data through observation enables the researcher to monitor and see how a product is used or a situation is handled in the real world. While the researcher is watching how individuals react in everyday occurrences, he gains valuable insight. One of the main strengths of observation is that the data being collected does not depend on the impressions or knowledge of the individuals being observed, but rather on the interpretation of the researcher of the action they are performing. Hence, the researcher is not contingent on what the participant has done in the past or claims to have done, does not have to deal with credibility issues, and memory lapses do not influence the quality of the information gathered.

There are a number of parameters that the researcher controls when starting an observation activity. He can declare the observation activity to the participants (the participants are made aware that they are being observed) or he can do it discreetly (the participants do not know they are being observed). The advantage of hiding his observation activity is that he will not influence and modify the normal course of the activity, but the downside is that he loses the opportunity to ask questions as events occur. He can also do observation in a natural setting (the participants are in their normal environment) or in a laboratory setting (the market researcher sets up a scenario and then observes the actions of the participants). Finally, the researcher can choose to directly participate in the activity to increase his comprehension of the phenomena. Participation observation allows interaction while recorder observation is easier to analyze. For example, a market researcher could work as a volunteer in a hospital to better understand the patient's waiting experience, chatting with patients who have been waiting for a while. In this case, the observer has to be careful not to "contaminate" the observation experience and influence responses, and has to be mindful of potential bias: it is quite easy to become attached to individuals, thus biasing the evaluation process. Finally, the presence of the participant–observer can change normal dysfunctional patterns or functions, and the researcher should account for the impact that his presence will have on normal processes.

Coming back to the earlier example on advanced insulation, my client was developing a very robust insulation container, and wanted to see if it could be used by fishermen. To understand how fishermen use existing insulation containers in their everyday setting, I went to a dock and observed fishermen coming back from fishing trips. The objective was to record and observe how the fishermen handled their insulation containers after a day at sea. The observations were surprising: in some situations, the fishermen would simply throw their containers overboard when arriving at the docks from the sea. I also noticed that quite a number of containers were pierced by the forklift carriers. As such, while in literature, competitors claimed that their products has a useful life of 10 years, our observation at the dock revealed that everyday usage was brutal and incompatible with the product my client was developing.

One key difference between my client's product and traditional insulation is that a normal insulated container can be pierced by a forklift, "patched" with insulation foam and some duct tape, and then sent back to the docks. It would still have an estimated 75%–80% efficiency rating, whereas my client's technology required 100% product integrity. The industry's definition of functional included heavily damaged containers, a fact that observation brought to light (Figure 2.1).

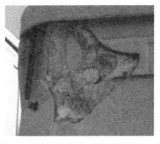 

Figure 2.1 **A few containers damaged but still used in a day-to-day situation by fishermen.**

There are some conditions to using the observation method. When choosing this methodology, the researcher should make sure that the information is observable, or inferable: choosing to observe a customer's purchasing decision process wouldn't be possible, but observing how they use a product is possible. Also, the phenomena to observe must be frequent, repetitive, or predictable. If not, it becomes too costly to observe. Finally, the observation must be relatively short in duration to be cost-effective.

As mentioned earlier, observation has a number of strengths over other tools. First, it allows direct access to the person being observed without the use of intermediaries. As the researcher is observing, there are fewer opportunities for the subject to respond dishonestly. Finally, it allows the researcher to observe the subject in his normal environment, which allows him to notice elements that the subject himself is unaware of.

The disadvantages are also pretty steep. First, as the observer is collecting the information, he is inserting himself as a part of the data collection equation. His point of view, past experiences, and even mental state (fatigue, emotions) can all taint the information he is collecting. A researcher observing a setting that he has himself experienced in the past can have preconceived expectations that he specifically looks for. Also, the presence of the observer has a direct impact on the subjects he is observing, especially if they are aware of his presence and are constantly questioned about decisions they make or actions they take. Finally, there is an undeniable cost to performing observation, both financially and timewise. Collecting significant observational data can take a lot of time on the part of the researcher.

2.5.6.1 Use in Life Sciences

Observation can be used in a variety of situations. For example, it can be used in environments with patients, observing how medical devices are being used in the healthcare environment, or by a patient in a home setting. It can also be useful to learn how patients are interacting with doctors in certain situations.

One of the emerging and interesting variations in observation is the use of wearable technologies as a means to observe and understand true patient behavior. Wearable technologies are devices that are worn by individuals, and that monitor the person's movement, heart rate, sleeping pattern, and more. There are many different devices that can be worn on wrists, as vests, on footwear, and more. Contract research organizations (CROs) are already using them as a way to monitor participants in research studies, and the U.S. Food and Drug Administration has issued guidance presenting regulatory views on the use of mHealth technologies in clinical trials. The increasing prevalence and familiarity of these devices among consumers, as well as the increase in accuracy, has attracted interest from many of the big players in this space.

There is also a relevant trend in the use of the participant's own mobile device in clinical trials, called Bring Your Own Device (BYOD) studies. As of 2016, Kara Dennis, managing director of Mobile Health (mHealth) at Medidata, stated that the concept of BYOD has reached "an inflection point," noting that one-third of the company's mobile health trials include a BYOD or hybrid BYOD component. In these trials, subjects can use their own smartphone or tablet to complete study-related tasks or if they do not have a qualifying device, they can be provided with one.* Benefits to BYODs include lower training costs and lower setup costs (users are already familiar with their devices), but there is the concern over Big Brother anxiety (as some users find continuous monitoring unsettling). Besides, the Scanadu story from 2016 (where users who purchased a device for a clinical study were not told it would be

---

* EyeforPharma. 2016. Patients at the helm: Putting patients in charge of clinical trials. Trends in Patient-Led Clinical Trials, http://www.mediantechnologies.com/wp-content/uploads/2016/12/eyeforpharma-Trends-in-Patient-Led-Clinical-Trials.pdf (Accessed January 19, 2017).

shut down after a predetermined date to comply with FDA regulations) might stop the whole BYOD space in its tracks, especially in situations where consumers have a vested interest in the device.*

Wearable technologies automate data collection as the researcher does not need to question participants. He can simply monitor their behavior through the device. They also produce much more precise data as participants do not have to estimate answers and there are no issues relating to recalling answers; the researcher has access to the exact data precisely calculated by the device, which he can obtain through an Internet connection. As such, this method mitigates one of the key disadvantages of observation, which is the time necessary to observe participants.

The HealthKit Wellness App developed by Apple is an interesting tool dedicated to wearables; it consolidates health data from iPhone, Apple Watch, and third-party apps, turning the phone into a patient-monitoring tool. It can be programmed to monitor a number of metrics, such as heartbeat and movement, and can send automated messages once a predetermined condition is reached.

The increasing use of consumers' wearable fitness trackers can also be a boon for companies wishing to understand consumer patterns. And there's no telling what the future holds in this space. Imagine using Google Glasses (a head-mounted display shaped in the form of glasses that can record activity) to monitor a person's shopping experience, and later analyzing shopping patterns. Or feasibly, a company could use 3D virtual tools to enhance focus group participants' experiences by simulating environments. The emergence of and rapid changes in technology will play a role in how market research observation occurs, and how data can be collected.

---

* Robbins R. 2016. Customers flocked to this futuristic medical tricorder. Now they're crying foul. *Stat.* https://www.statnews.com/2016/12/14/scanadu-tricorder-crowdfund/ (Accessed January 19, 2017).

2.5.7 Mystery Shopping

Mystery shopping is a market research activity where trained individuals are mandated to evaluate the quality of service or the compliance with regulation. To do so, these individuals are sent to experience the organization from the outside, as an everyday customer. When performed, the locations that are the subject of the mystery shopping activity are not aware that they are being evaluated. This allows the head organization to better appreciate how it is performing, and to adjust processes and internal procedures accordingly.

The mystery shoppers act as much as possible like everyday consumers, asking questions, purchasing products, or behaving following certain scenarios, and then reporting their experiences back to the market researcher coordinating the effort. Mystery shoppers can investigate a number of items such as

- The customer experience (number of employees, ease of service, attitude of employees)
- The facility (cleanliness, ease of navigation)
- The purchasing experience (time spent to purchase product, ease of transaction)
- The post-purchasing experience (ease of returning product or using the return policy)

While it is possible to do mystery shopping in-house, it is more efficient to hire third-party companies or freelancers to engage in mystery shopping. They are more likely to reflect the point of view of an unattached observer, unlikely to nuance their observations with their own insider knowledge, and less likely to be recognized by staffers being evaluated.

Finally, it is possible to use mystery shopping to evaluate competitors, but a certain number of ethical guidelines should be respected. First, the mystery shopper should not record competitors' employees with a recording device as these employees have not given their consent. Also, the length of time spent should be the equivalent of a normal market transaction (so as to not waste competitors' resources unethically). Finally, the evaluation scenario should not require a follow-up call from the company being observed, unless this is a normal part of this transaction. Globally, if the mystery shopper is asking questions that a shopper usually asks (e.g., price, quality, or availability of goods) and if no confidential business information is sought or revealed, then the common sense rule of "no harm, no foul" applies.* Even then, if these guidelines are respected, some experts believe that mystery shopping competitors is walking into a gray area of market research, putting you at risk of liability.

2.5.7.1 Use in Life Sciences

Mystery shopping is quite useful for mature organizations that are already generating sales (or offering a service), and need to gain a better understanding of their customer experiences. It is also quite popular in healthcare facilities, as well as in new healthcare mobile application technologies.

In mature organizations (generating sales), mystery shopping is useful to evaluate the customer process and identify pain points in the enquiry, purchasing, and maintaining phase. If you have issues generating recurring purchases, consider a mystery shopping initiative to evaluate your own internal processes.

Healthcare facilities use mystery shopping to evaluate the level of service they are offering. It provides insight into the patient's experience, and identifies gaps in the service chain. As such, a typical healthcare mystery shop could include

---

* Craig E. 2007. Mystery shopping. *Competitive Intelligence Magazine*, Vol 10, Number 3, May–June 2007.

1. Scheduling an appointment with a healthcare provider
2. Visiting the healthcare center
3. Completing a medical consultation
4. Reporting back on the patient experience

Also, mystery shops are often used by technology companies developing digital applications. For example, some firms will engage mystery shoppers to download, install, and register their app and then report on ease of use, issues during installation, and overall feedback.

Finally, some telemedicine apps will use mystery shoppers in an effort to evaluate the application's ease of use and solidity. They will use mystery shoppers to engage with doctors, enquire about medical conditions, and even have the mystery shoppers use the app to purchase medications (OTC and prescription) online. Typically, various elements are being verified, from the customer experience to quality of service all the way to adherence to regulation (is the customer able to purchase restricted products, or is regulatory compliance enforced?).

2.6 Closing Words on Ethics and Primary Market Research

Throughout this chapter, we presented and discussed many different tools that a market researcher can use to collect data from participants. Sometimes, these participants will be aware of the marketing research effort, sometimes not. In any case, there are a number of ethical considerations that a market researcher must remember.

First, it's very important to be honest when you collect data, identifying yourself and describing why you are collecting the data. Misrepresentation is a huge issue in data collection, and it is tempting to do so in an attempt to ease the data-gathering process. For example, some researchers will pretend to be a potentially interested party, or pose as a student gathering information for a school project. This is clearly unethical. Instead, efforts should be spent identifying targets that have the information and are more likely to want to share it, such as academics, technology vendors, advertising agencies, and journalists, for example.

Also, if the participant asks, you should be able to answer what the project is for. Not being able to give answers on the nature of the project gives the appearance of spying, not market researching. Companies engage in market research activities all the time, and you should be able to answer these types of questions truthfully. If you cannot, then ask yourself if this is a project that you should be doing.

It is very easy for a market researcher to influence the participant's response. Asking leading questions can cause a participant to answer in a specific way. Even agreeing with a participant rather than impartially acknowledging their answer can influence the participation, tainting future responses, and fundamentally making it unusable. One has to remember that leading participants to answer a certain way might let you find the answer you want to hear, but this will not reflect the real market's appreciation of your product: wouldn't you rather find out the real market requirements during the market research phase, rather than shaping market research to fit your preconceived ideas, and then failing during commercialization? *While market research shapes our vision of the market, it does not change the true nature of the market.*

Furthermore, respect the confidentiality of the participants. If you have given them the assurance that you will protect their responses, be ready to do so. If you believe you cannot assure the participants' confidentiality, or if you do not intend to do so (by sharing the results to other stakeholders), be upfront so that participants can have the option of opting out. As an example, a few years ago, a client had contacted me to conduct a survey of life science companies. Participants were assured anonymity, and answered truthfully about sensible topics such as cash on hand, burn rate, and strategic priorities. After compiling the data, the client became interested in some of the information collected, and wanted access to granular data so that he could share it with other stakeholders. I refused. While I lost a client, I maintained the privacy I had assured participants.

Finally, *primary market research is not a commercialization activity*. Engaging participants in market research and then trying to sell a product midway during data collection undermines market research as a whole. In some cases, a client might express interest in a product you are researching. When this happens, my approach is to ask the participants "You seem to have some interest in product X. Would you like me to refer you directly to Company X as an interested party? Do you accept that your coordinates be shared directly with the appropriate person?" As such, the participant is authorizing you to share his information and his interest. You are serving both parties, and with consent, are relieved of other obligations you might have (such as keeping the data anonymous).

Chapter 3

Secondary Research

Secondary research is the collection and collation of information that is published and publicly available. Also called desk research, it is frequently done to explore a topic before engaging in more expensive primary research, or to quickly gain a summary understanding of a topic without engaging too many resources.

The advantages of doing secondary research (vs. primary research) are numerous. First, secondary research is considerably less expensive than primary research. Also, it is considerably faster to complete as it does not depend on third parties (such as recruited participants, organized focus groups, or enrolled survey participants) to obtain the information, and the sample size of third-party research reports will often be quite considerable. Furthermore, due to scope and reputation, third-party information often has more perceived authority and impartiality than "in-house" research. An assessment such as "BCC Research estimates that the global advanced drug delivery market should grow from roughly $178.8 billion in 2015 to nearly $227.3 billion by 2020, with a compound annual growth rate (CAGR) of 4.9%"* has more credibility than most in-house estimates. Finally, as mentioned at the beginning of this section, secondary research is very useful to orient and define primary research: a researcher will often start a market research project by doing a quick market review to identify some of the main trends and concerns before diving into primary research.

There are some disadvantages to secondary research. First, the information is not always personalized to an organization's requirements, and it is quite difficult to find data for emerging fields: reports on nanomedicine are plentiful, but developing a specific application merging nanomedicine and information technology means that secondary research will be quite scarce, and means that the organization will either need to (a) extrapolate from generalist research, or (b) conduct primary research. Also, the data might be outdated, limiting its usefulness. Finally, most of the time, the original data used in secondary data is unverifiable: it is quite difficult to spot errors in data collection or dispute the way that data was analyzed in a consolidated report.

There are two types of secondary data: external and internal. External secondary data is information gathered from outside the organization. This includes anything from government statistics to media sources. Internal secondary data is data that the organization is generating itself. It could be data collected from customers' feedback, accounting and sales records, or employee experiences. While it is possible to do secondary research using non-Internet sources, the bulk of our suggestions are related to this medium for both the ease of use, convenience as well as cost.

Finally, we make a distinction between active and passive secondary research. Active secondary research takes place when the researcher is actively searching for information,

---

* Wadha H. 2016. Global Markets and Technologies for Advanced Drug Delivery Systems. http://www.bccresearch.com/market-research/pharmaceuticals/advanced-drug-delivery-systems-tech-markets-report-phm006k.html (Accessed January 2, 2016).

while passive secondary research is the use of tools and software to automate data collection.

3.1 Active Secondary Research

Active secondary research takes place when the market researcher is dynamically searching for information. In this section, we will be going over the most popular and pertinent sources of data available. All of the suggested sources are web based. The Internet has grown to be by far the most important resource for searching for and identifying secondary research information: even most magazines and newspapers have made their archives (partially) available online.

While there are some ethical considerations for what information the researcher can use, a simple rule of thumb is that any information made available to the public is fair game for collection and review. If the method used to obtain the information is not commonly available to the public (such as using a former employee's password to access a restricted website area), then it is not only unethical, but most definitely illegal.

In the following section, we will be going over a variety of Internet data sources that a market researcher might wish to investigate when doing secondary research, and will conclude with a few tips to increase the effectiveness of Google-based Internet researches.

3.1.1 Government Data

Government agencies generate large bodies of information that can be used by researchers. Most of this information is free to use, and can be useful at the start of a research project. Government data is usually statistical in nature, and is very useful when building marketing models or trying to understand the nature of a market. Table 3.1 provides a few

Table 3.1   Government Databases for Secondary Research

Source	Website
CIA World Fact Book (Country profiles)	www.cia.gov/library/publications/the-world-factbook
Global Health Observatory (World Health Organization)	www.portal.pmnch.org/gho/en/
World Bank Open Data	data.worldbank.org/
FDA Medical Device Databases	www.fda.gov/MedicalDevices/DeviceRegulationandGuidance/Databases/
American Fact Finder (U.S. Census Bureau)	www.factfinder.census.gov
U.S. Patent Search	patft.uspto.gov/
National Center for Health Statistics	www.cdc.gov/nchs/hus.htm
National Agricultural Statistics Service	www.nass.usda.gov
U.S. Census	www.census.gov
Data.gov (U.S. Government's open data)	www.data.gov
FastStats (Centers for Disease Control and Prevention)	www.cdc.gov/nchs/fastats/Default.htm
Statistics Canada Portal	www.statcan.gc.ca
Eurostat (European Union)	ec.europa.eu/eurostat
EU clinical trial portal and database	eudract.ema.europa.eu
ASEAN Statistics	asean.org/resource/statistics/asean-statistics/

government databases that could be useful to review during a market research effort in life sciences.

3.1.2 Public Company Data

Companies publish a lot of useful information online. Often, companies will do market research, and will publish results as consolidated information in corporate documentation. As such, reviewing this publicly available data is another useful way to start your research effort.

One of the first places to search online is the website of a specific company with a technology or product akin to the one that is the focus of your research. Looking at data from other companies can help to estimate a number of elements, including market size, growth, and segmentation, since you will possibly find pieces of market data. For example, while perusing a corporate presentation, you might find a sentence on market segmentation where the company states that it believes to be the second largest player in the market, and estimates that it generates 30% of the total sales. This statement might be based on other secondary research or on internal data, so it's important to note the origin of the statement. Some of the documents that can be reviewed online include

- Annual reports, which might also be available on the websites of central depositories, such as the Securities and Exchange Commission (for American companies, www.sec.gov) or SEDAR (for Canadian companies, www.sedar.com)*
- Company pitch decks or presentations at investor's forums

---

* Unfortunately, there aren't any consolidated sources for European and Asian companies, so you will have to search through local exchanges for the information you need. The first step will be to identify where the company is listed.

■ Press releases, marketing collateral, product brochures, and white papers

■ Video product presentations (take the time to visit YouTube to verify if the company has a YouTube channel, and if it has published online content on it)

Once these documents are identified, be sure to download them and save them to your own local drive. Companies can change or update their content online without warning, and the information you identified might not be available any longer when you return to the website. Taking a screenshot of important information can be another way to ensure that the data isn't lost following a website update.

3.1.3 Print Media

Print media sources are published on a regular schedule by specialized companies. These include magazines and newspapers, and include both their printed/physical format as well as their Internet counterparts.

Generalist newspapers and publications, while interesting, are of limited use when doing specialized market research. Although it is possible to get general information from them (such as trends, government policies and orientations, or industry summaries), print media market research in life science will focus mostly on specialized content publishers, mostly in the form of trade magazine publishers.

There are a number of trade magazines that are published on a regular basis that are of use to market researchers. Most of them are free, and access to their archives is public most of the time, although some do monetize their archives. Table 3.2 lists some of the main publications that market researchers can use during their market research efforts. One thing to investigate is if the publication publishes specialized inserts or special issues on focused topics, as these also provide very valuable market data.

Table 3.2   Publications of Interest in Trade Media for Life Sciences

Publication (Website)	Description
Genetic Engineering & Biotechnology News (www.genengnews.com)	Bioindustry news publication
FierceBiotech (www.fiercebiotech.com)	Biotechnology news publication
Fierce Pharma (www.fiercepharma.com)	Pharmaceutical news publication
BioProcess International Magazine (www.bioprocessintl.com)	Manufacture of biotherapeutics
Contract Pharma (www.contractpharma.com)	Pharma outsourcing industry
MD+DI (www.mddionline.com)	Medical device and diagnostic industry

3.1.4 Trade and Industry Groups

Trade and industry groups are organizations representing multiple firms (private companies, government agencies, universities, consultants) in a common commercial activity sector. Some of the larger associations produce (or sponsor) reports relevant to their industry, which can be of use during market research. One of the advantages of the reports produced by these organizations is that they are third-party reports and are (relatively) unbiased, but the reports are not always freely available, and are sometimes reserved for members.

Associations are also useful to identify participants or targets for further research as their members are routinely listed on their website. This can give the researcher a good starting point when researching a specific technology area, or if the researcher's research focus is in a specific geographical region. Finally, many trade associations collect information from their members in the form of surveys and then share the results at large in a consolidated format, producing reports on key trends and issues of concern. The most important life science trade organizations are listed in Table 3.3.

Table 3.3   Main Life Science Industry Groups

Source (Website)	Focus
Biotechnology Innovation Organization (www.bio.org)	Innovative healthcare, agricultural, industrial, and environmental biotechnology products
PhRMA (www.phrma.org)	Innovative biopharmaceutical research companies
Generic Pharmaceutical Association (www.gphaonline.org)	Manufacturers and distributors of generic prescription drugs
American Hospital Association (www.aha.org)	Represents hospitals, healthcare networks, and their patients
European Hospital and Healthcare Federation (www.hope.be)	National public and private hospitals
MedTech Europe (www.medtecheurope.org)	Alliance of European medical technology industry associations
Medical Device Manufacturers Association (www.medicaldevices.org)	Innovative and entrepreneurial medical technology companies

3.1.5 Scientific Publications

Scientific publications are a very useful source of information when researching market trends and future technologies. They can also be very useful to identify key opinion leaders for future research efforts, such as interviews and focus groups, and they supply valuable information to benchmark technologies. While there are many publications available, the best approach is to target hubs that consolidate articles (Table 3.4).

Table 3.4   Main Scientific Article Hubs for Market Research

Source (Website)	Focus
PubMed (www.ncbi.nlm.nih.gov/pubmed)	PubMed comprises more than 26 million citations for biomedical literature
BioMed Central (www.biomedcentral.com)	Peer-reviewed biomedical research specializing in open access journal publication
Science Gateway (sciencegateway.org/)	Directory of links to biomedical science and research journals
Europe PMC (europepmc.org)	Europe PMC also contains patents, NHS (National Health Service) guidelines, and Agricola records
Google Scholar (scholar.google.com)	Freely accessible web search engine that indexes the full text or metadata of scholarly literature

3.1.6 Market Research Firms

Large market research firms publish syndicated reports on a regular basis. They collect data from major participants in an industry, consolidating the data and publishing it so that it can be purchased by all interested parties. The reports can be expensive, but can be worth it for the time saved by purchasing a completed product (rather than engaging in a costly primary research operation). Some companies also offer report customization for its bigger clients, enabling some level of personalization. Finally, executive reports on these syndicated reports are sometimes available at large, so it is possible to collect some data that can be used for data triangulation.

3.1.7 Competitive Start-Up Research through Specialized Websites

Researching upcoming competitors or new technologies can be accomplished by searching specialized websites. There are a number of specialized databases that can be consulted for free that are of interest to emerging start-ups to identify early competing technologies or partnership opportunities (Table 3.5).

On the importance of competitive intelligence ...

As an investor, we talk to many companies trying to treat the same therapeutic areas. Usually a company that can pitch by focusing on the strengths of its own programs—rather than criticizing its competitors—tells a more valuable and compelling story.

Caroline Stout, Investor at EcoR1 Capital

Table 3.5 Specialized Databases for Start-up Competitive Intelligence

Database (Website)	Description
CrunchBase (www.crunchbase.com)	Database of innovative companies and entrepreneurs
YouNoodle (www.younoodle.com)	Platform for entrepreneurship competitions
AngelList (angel.co)	Website for start-ups, angel investors, and job-seekers looking to work at start-ups
BizStats (www.bizstats.com)	Business statistics and financial ratios

3.1.8 Blogs

Blogs are websites or publication streams maintained by individuals and companies to publish information in a more interactive format. The objective is to generate discussion between the company and its consumers, or between an expert and people interested in a topic. It is also heavily used by consumers and advocacy groups to express opinions, both positive and negative, related to a topic of interest.

Blogs are one of the top trusted sources of information due to their personal nature. A Technorati survey found blogs to be one of the top five trusted sources of information on the web (after news sites, Facebook, retail sites, and YouTube).* But the fact that they are so trusted has led to a deterioration in the quality of information being shared on them. It is quite easy for bloggers to manipulate (or be manipulated), so the information they share should be taken as an information lead, rather than being positioned as a definitive fact. As Ryan Holiday mentions in his book *Confessions of a Media Manipulator*, "The economics of the Internet has created a twisted set of incentives that make traffic (to the blog) more important—and more profitable—than the truth."†

Also, the popularity of blogs has been declining, but they still remain an important information source. While many people debate the usefulness of blogs when so many ways to push information to consumers exist, blogs still have some very important characteristics compared with alternatives (e.g., social media websites such as LinkedIn or using a Facebook Feed). The most important of these differences is that as blogs can be located directly on an organization's website, they allow for more detailed posts and have more longevity, as they

---

* Technorati Media. 2013. Digital influence report. http://technorati.com/wp-content/uploads/2013/06/tm2013DIR1.pdf (Accessed December 16, 2016).
† "Trust Me, I'm Lying: Confessions of a Media Manipulator" (Portfolio Penguin, 2013), written by Ryan Holiday, is an interesting read on the economics and role of the Internet in how information is generated in social media.

are available for as long as the organization wishes them to be online.

Researching blogs is a useful way to identify trends, consumer reactions, and key opinion leaders. Blogs from advocacy groups are especially useful to identify concerns from patients, and blogs from key opinion leaders are useful to identify upcoming trends, but care should be taken to measure these opinions against other sources of information.

3.1.9 Social Networks

For the purpose of this book, "social networks" are defined as dedicated websites or applications where users with a common objective gather and participate in discussions. These discussions create a network of social interactions, as users share messages, comments, information, opinions, experiences, and more.

As of 2016, the most popular social networks are LinkedIn (which is mainly used by the business community), Facebook (which focuses more on social interactions), and Twitter (which is used for commenting and quickly diffusing information and opinions). However, there are hundreds of social networks catering to specialized needs for niche communities. A very interesting one for researching trends in life sciences is the PatientsLikeMe (www.patientslikeme.com) social network, a website that bills itself as a "free website where people can share their health data to track their progress, help others, and change medicine for good."

Social networks are a useful way to measure the customer's pulse on a topic. Analyzing the comments found on social network pages allows the researcher to understand how people view a brand, or their perception of a specific topic. It is also possible to use social networks to recruit participants for surveys and focus groups, as well as interacting directly with them and engaging in one-to-one conversations.

The popularity of social networks has led many companies to set up and maintain pages on social networks as well. Some

pages are for the company in general, while others are specific product pages. When reviewing these pages, it is possible to collect and collate customer feedback on these specific products.

For example, say you were developing a carrying device for glucose monitoring devices. Using Facebook, you would find the Medtronic Diabetes Facebook page.* Reading through the comments, you would find information such as:

✓ *I would like to know why the pump clips are made so cheaply now* ☹ ☹ *! I am considering gluing it to my pump at this point because I am so tired of it snapping off with the slightest bit of a bump and no it's not a default in my clip itself I had two sent to me and both are just as bad as one another ... the quality of this product is going down but the prices seem to keep going up and it is truly sad ...*
✓ *Both of my clips have broken in the last few months. One of them broke while I was sleeping!*
✓ *I stopped using those plastic ones for that reason*

Through more research, you might be able to identify an opportunity for a more sturdy carrying solution, or something completely different from the current technology.

A good model for analyzing data gathered on social media is using a slight variation of the Five Ws (Who / What / Where / When / Why).

✓ **What happened?**—What was the situation that caused the posters to post messages on social media?
✓ **Who is involved?**—Who is posting the message? Who else is involved? Is the poster posting for somebody else?
✓ **Where was the comment posted?**—Where did the poster post his comment? Did he append his comment to a specific post, or did he start a new conversation?

---

* Source: www.facebook.com/medtronicdiabetes (Accessed December 25, 2016).

✓ **Why did that happen?**—Why did he post a message? Why did he use social media instead of a more conventional method (such as calling customer support)?
✓ **How can this problem be solved?**—What did he hope to accomplish? What are the solutions that would resolve this situation?

There are also many online tools that a market researcher can use to monitor customer feedback instead of using a generalist search engine. For Twitter, there are websites such as MentionMapp (http://mentionmapp.com/), which lets you create a map of users and hashtags based on your search, enabling you to identify social key opinion leaders. Another option is TweetReach (https://tweetreach.com/), an analytical tool that generates real-time stats for any search term by searching through Facebook, Twitter, and Instagram. Finally, Mention (https://mention.com/en/) does real-time monitoring of both the web and social media, enabling you to do competitive analysis and to find influencers while Social Searcher (https://www.social-searcher.com) is a free social media search engine that looks through available public information in many social networks (such as Twitter, Google+, Facebook, YouTube, Instagram, Tumblr, Reddit, Flickr, Dailymotion, and Vimeo) for the information you need, while also allowing users to save their searches and set up e-mail alerts.

3.1.10 Discussion Boards

Discussion boards are online discussion forums where users can discuss topics through forum posted messages, often anonymously. Their main strength is that the messages are usually more detailed than those found in blog post responses, and have longer archival periods. One of the most useful discussion groups for life sciences is cafepharma.com, where you can find great insights on pharma marketing trends and rumors. Reddit is another good source for finding informal information.

For example, on Reddit in November 2016, someone started a discussion called "What are your thoughts on the Bayer x Monsanto merger?"* Some of the feedback included:

■ *Could be something to do with weed legalization. I know Monsanto has wanted to grow large scale pot farms, and that means a lot of drugs to process. Bayer has a history of processing drugs especially for the pain relief sector. I have no proof of this whatsoever. But no one else commented and as a weed smoker this merger turned some lights on for me.*
■ *That conspiracy theory just never dies. They have stated (over and over and over) that they have no interest in marijuana.*
■ *Fun thought, but actually extremely improbable bumping up against impossible. By this logic, why not get into the tobacco market? Growing commodity crops, which is what marijuana would be if legal, is not in the interest of a chemical company. Bayer wants to get into agricultural chemicals. More so than they already have and the easiest way to achieve that when you're sitting on mountains of cash are to buy someone.*

Let's be clear, these are rumors, not facts, akin to something a researcher might overhear in a café or a convention. And without data triangulation, these are random statements, and shouldn't be taken at face value. Nonetheless, what they do provide is a potential clue, and a research lead.

These two websites are examples of places where people familiar with the industry converge to engage in discussions. It is possible for a researcher to use these to test a hypothesis, or to discreetly ask participants a question to get more potential information.

---

* Reddit Forum. 2016. "What are your thoughts on the Bayer x Monsanto merger?" https://www.reddit.com/r/biotech/comments/5i8x5a/what_are_your_thoughts_on_the_bayer_x_monsanto/ (Accessed December 27, 2016).

3.1.11 Search Engines: Tips and Tricks

Searching through the Internet using a web research engine is usually the first step of secondary research. Here are a few tips and tricks to make the research effort more efficient:

■ *Look to the past*: Sometimes, a market researcher will be looking for something specific, but will conclude that the information is no longer available online. It might be an old press release that a competitor has pulled from their website, information on a previous partnership that was quietly ended, or specifications on discontinued products, for example. In these cases, the website www.archive.org (an Internet archive non-profit digital library offering free universal access to all) is especially useful. Archived in their public database are historical web snapshots of a company's website, which can include pages, attachments, and more. While a researcher might not have access to each version of a company's website, there are often several snapshots taken throughout the year, enabling the researcher to identify key information that has been removed online.

■ *Look for corporate web DNA*: I originally found this technique referenced by Leonard Fuld in *The Secret Language of Competitive Intelligence*. It is based on the concept that every organization develops their own brand of corporate speak, or pattern. It is akin to corporate web DNA. Fuld defines it as "a unique pattern of words and phrases that form the substance of a company's website, its press releases, and its advertisements."* As such, if the researcher can identify a group of unique words or jargon as potential corporate DNA, he can then proceed with searching the web using the aforementioned terminology, grouped between two sets

---

* Fuld L. 2006. *The Secret Language of Competitive Intelligence*. New York: Crown Business.

of quotation marks. As an example, using Medtronic's "Transforming technology to change lives" slogan to search the web brings up a series of white papers (old and new), job offers (both current and expired), as well as customer testimonials.

■ *Go beyond Google*: If you are not finding the information you need, you can try using another search engine to obtain different search results. Some of the interesting alternative search engines include
  – Bing (www.bing.com) (which is reported to have a better video search option)
  – Board reader (boardreader.com) (which specializes in user point of views by searching through forums, message boards, and Reddit)
  – Slide Share (www.slideshare.net) (a cornucopia of PowerPoint presentations, slide decks, and webinars from past conferences)

3.1.12 Power Up Google

Google is the starting point for many research projects. Google accounts for over 75% of the global desktop search engine market share, as its simplicity of use and search algorithm for ranking websites in order of importance enables researchers to quickly find the information they are looking for.

When searching on Google, remember that the search engine ranks the first word as slightly more important than the second, and the second slightly more important than the third, and so on. Hence, searching for "north America vaccine trends" and "vaccine trends north America" will generate slightly different results (Table 3.6).*

---

* Search demonstration completed January 9, 2017.

Table 3.6 Demonstration of the Importance of the Order of Words When Searching Google

Search Item—North America Vaccine Trends	Search Item—Vaccine Trends North America
Porcine Vaccine Market, and Porcine Circovirus ... - PR Newswire www.prnewswire.com/.../porcine-vaccine-market-and-porcine-circovirus-associated-d... ▾ 5 days ago - Porcine Vaccine Market, and Porcine Circovirus Associated Disease: End User - North America Industry Analysis. Size. Share. Growth. Trends. ...	Porcine Vaccine Market, and Porcine Circovirus Associated Disease ... www.prnewswire.com/.../porcine-vaccine-market-and-porcine-circovirus-associated-d... ▾ 5 days ago - Porcine Vaccine Market, and Porcine Circovirus Associated Disease: End User - North America Industry Analysis. Size. Share. Growth. Trends. ...
North America Porcine Vaccine Market to US$926.2 Million by 2024 ... www.fox34.com/.../north-america-porcine-vaccine-market-to-us$262-million-by-2024-... Nov 29, 2016 - The North America Porcine Vaccines Market for features an essentially ... North America Industry Analysis. Size. Share. Growth. Trends. and ...	North America Porcine Vaccine Market to US$926.2 Million by 2024 ... www.prnewswire.com/.../north-america-porcine-vaccine-market-to-us$262-million-b... ▾ Nov 29, 2016 - The North America Porcine Vaccines Market for features an essentially ... North America Industry Analysis. Size. Share. Growth. Trends. and ...
Current Trends Measles -- North America, 1984 - CDC https://www.cdc.gov/mmwr/preview/mmwrhtml/00000664.htm ▾ Current Trends Measles -- North America. 1984. Since the introduction of measles vaccine in North America. Canada. Mexico. and the United States have made ...	ᴾᴰᶠvaccine market place - World Health Organization who.int/influenza_vaccines_plan/resources/session_10_kaddar.pdf ▾ NEW TRENDS SINCE 2000 ? . IMPLICATIONS ? . North – South GAP . Influenza vaccine market estimated at $2.9 billion in 2011 to $3.8 billion by 2018
Images for north america vaccine trends RESEARCH → More images for north america vaccine trends	Report images	Vaccination Trends | FluVaxView | Seasonal Influenza (Flu) | CDC www.cdc.gov/flu/fluvaxview/trends.htm ▾ Sep 29, 2016 - FluVaxView vaccination trends. CDC . Influenza Vaccination Reports. Trends. Influenza Seasons 2007-2008 through 2010-2011.
Forecasted trends in vaccination coverage and correlations with ... thelancet.com/journals/langlo/article-PIIS2214-109X(16)30167-X/fulltext Aug 25, 2016 - Forecasted trends in vaccination coverage and correlations with ... More generally Europe. North and Central America. and much of Asia ...	Vaccine - Wikipedia https://en.wikipedia.org/wiki/Vaccine ▾ A vaccine is a biological preparation that provides active immunity to a particular ... Early smallpox in North America . ... Factors that have an impact on the trends of vaccine development include progress in translatory medicine. ...
Vaccine - Wikipedia https://en.wikipedia.org/wiki/Vaccine ▾ A vaccine is a biological preparation that provides active immunity to a particular ... 1625. Early smallpox in North America . ... Factors that have an impact on the trends of vaccine development include progress in translatory medicine. ...	Global trends in vaccination coverage - The Lancet Global Health thelancet.com/journals/langlo/article-PIIS2214-109X(16)30185-1/fulltext Aug 25, 2016 - Global trends in vaccination coverage . regions of the world. this is not the case any more in Europe and. to a lesser extent. North America
	Digital Vaccine Labs (DVLabs) – TippingPoint - Trend Micro USA www.trendmicro.com/us/security-intelligence/research-and- digital-vaccine-labs/ ▾ Digital Vaccine filters help you gain control of your organization's patch management ... Threat Digital Vaccine (ThreatDV) . North America. +1 855 681 8324

Table 3.7   Tips for Searching with Google to Increase Results

Tips	Examples
Use the *similar* option under a search result to have Google Search look for similar web sources.	Rofecoxib (Vioxx) voluntarily withdrawn from market www.cmaj.ca/content/171/9/1027.full ▾ by B Sibbald - 2004 - Cited by 40 - Relat [Similar] Oct 26, 2004 - Merck & Co. announced ...untary worldwide wit after a study showed patients taking the drug on a ...
The "~" punctuation mark makes Google search for synonyms to a specific word	Searching for ~neuroscience will search for neuroscience as well as neurophysiology, neurobiology, brain, and neurology
Include words in your search by using the *addition sign* before words or phrases, or exclude words from your search phrase by using the *minus sign*	Searching for "biosimilars +Asia" will bring back every search result with biosimilars and Asia. Searching for "biosimilars –Europe" will bring back every search result with biosimilars that excludes Europe
Restrict your search to a specific domain by using the *site* command; to get results from multiple sites or domains, combine with the *OR* operator	To find information on Aspirin published specifically in *The New England Journal of Medicine* or from the FDA, you would type: *Aspirin site:eejm.org OR site:fda.gov*
The *filetype* operator lets Google search specifically for exact types of files, which is quite useful when looking for market data	To find some PowerPoint slides with information on market share for cardiology devices, you would type: *Cardiology devices "market share" filetype:ppt*

There are a number of powerful options available when searching with Google.* Some of the most useful ones are shown in Table 3.7.

---

* Looking for information on Google is quite complex. Fortunately, Google offers a series of educational course online for free for those who wish to learn more. You can see these courses by going to http://www.powersearchingwithgoogle.com/ (Accessed December 25, 2016).

3.2 Passive Secondary Research

Automated Internet research tools are a boon to market researchers, as they automatically monitor and report on specific information topics. We will be going through some of the most interesting tools that researchers can use to automate their market research.

3.2.1 Rich Site Summary Feeds

Rich Site Summary (RSS) feeds are a simple method to aggregate data generated by specialized websites, and efficiently supply researchers with up-to-date information on specific topics.

Three types of RSS readers exist: web-based readers (which you access through your web browser), such as Feedly (www.feedly.com) and NetVibes (www.netvibes.com); client-based readers, which you download and install on your computer, such as RSSOwl (www.rssowl.org); and those that integrate into your web browser (Firefox and Internet Explorer both offer this option).

The advantages of using RSS feeds for research and continued monitoring are multiple. First, RSS feeds save time: you can quickly subscribe to the feeds you are interested in, and quickly scan aggregated data without having to visit every single website every time. Also, as RSS feeds update themselves automatically, you get information as it becomes available. Finally, they increase your productivity as you can quickly scan the headlines for the information you are interested in, and dig deeper on the topics of interest. Table 3.8 shows some RSS feeds you might want to follow.

3.2.2 Google Alerts

Google Alerts is a service offered by Google. It is useful to automatically monitor a topic by setting up a search alert.

Table 3.8   Top Life Science RSS Feeds

Feed	Topic
BioSpace.com—Featured news and stories	Biotech, clinical research, and pharmaceutical news
Food and Drug Administration—Press releases	Press releases from FDA
Medical Devices/Diagnostics News From Medical News Today	MedTech news
Healthcare Economist	Healthcare issues
Modern Healthcare Breaking News	News from modern healthcare
Forbes—Pharma & Healthcare	Pharma & healthcare stories

Once set up, the Google Alert sends search engine search results by e-mail to researchers as they occur, or on a predetermined basis in the form of a digest (once a day or once a week), at a predetermined time.

To create a Google Alert, the user only needs to go to the website www.google.com/alerts, type in their topic of interest, and customize the information feed requested (frequency, where the information will be collected from, the number of results wanted each period, and which e-mail will be receiving the information). It is possible to edit an alert if needed, or delete the alert if it is no longer needed.

Google Alerts monitors major sites and news media for the topics selected, and easily shares the information it collects across social media platforms such as Facebook, Twitter, and Google+.

3.2.3 Web Monitoring Tools

In addition to using automated alerting tools such as Google Alerts and monitoring RSS feeds, it is possible to set up specialized software and online search tools that monitor changes in specific web content (e.g., a competitor's website). These tools will alert you to any changes on a company's website as they occur, making it unnecessary to continually monitor key websites. Some software can also be configured to simultaneously monitor RSS feeds and newsgroups, making powerful all-in-one solutions. Some of the most popular software in this space includes

- Website-Watcher (www.aignes.com, which is software based)
- WatchThatPage (www.watchthatpage.com, which is free for individuals)
- InfoMinder (www.infominder.com, which is web based)

3.2.4 Social Media Tracking

A number of tools exist that are specifically designed to monitor social media. These tools search through popular medias, such as Twitter and Facebook, and use generated content such as blogs and comments to generate reports that can be used to identify underlying consumer trends.

One of the popular tools in this space is Keyhole (keyhole.co). This monitoring tool keeps an eye on keywords and hashtags across Twitter and Instagram and can be useful to quickly identify popular public key opinion leaders on a topic, which can then be engaged further for market research. It can also be used to identify geographic tends and estimate overall consumer sentiment.

Another useful tool is Warble (warble.co), which monitors Twitter and sends daily e-mail reports directly to the chosen e-mail account. It automates the process of monitoring Twitter, which is pretty important due to the significant mass of content that is generated on it daily. Twilert (www.twilert.com) is a variant of Warble, that while not free, enables the researcher to monitor words according to various vectors, such as only positive, negative, or neutral tweets. Hence, it already does the first step of the analysis for the market researcher.

3.2.5 Online Collaborative Tools: Factr

Factr (factr.com) is an interesting take on automated web monitoring, as it adds collaborative capabilities that allow team members to adjust the automated web monitoring tool and comment on the information being collected. Factr collects RSS feed data, which is fused into a single collaborative stream.

Factr allows teams to collaborate in real time while the platform collects data. As a user sets up a collaborative stream, data is continually collected. Interestingly, Factr has been online for a few years now, and some streams have been collecting data for over three years.

While searching through RSS feeds, Factr can also learn and make suggestions on new sources to collect data from. It can search for keywords, and users can upload images, files, and reports to the collaborative work space to increase collaboration.

Factr will also analyze tags, consolidate the research data, and generate reports, making it a useful tool to automate some of the more basic research.

3.2.6 Google Trends

Services such as Google Trends track comprehensive search results over long periods of time, and are useful to identify underlying trends. As such, a researcher can uncover what the potential target audience is searching for online and analyze the information to gain insights on potential customers (especially if the company is targeting end users).

Google trends will be useful to gauge overall interest in a topic by region, show the top queries related to the topic at hand, and try to identify related topics. It is a useful tool when identifying other relevant topics when researching a subject. Google Trends can be very useful to research a competitor and get a quick image of its overall popularity. Some other uses can include

■ Researching the product features that potential clients find important by identifying what search terms are the most popular in relation to the search topic.

■ Gauging consumer demand by searching which products or features generate the most consumer demand.

■ Evaluating what consumers are searching for in competing products, by entering a competitor's name in the topics box and looking at search queries related to the brand. For example, searching for Epipen in December 2016 generated related terms such as "epipcn copay," "epipen price hike," and "epipen price increase" as some of the top consumer searches.

3.2.7 A Word on Bots and Data Scrapers

One method of collecting data that might be suggested to the researcher is the use of "bots." Also called "scrapers," "robots," or "spiders," these are types of software that the researcher sets up to run automated tasks to collect data systematically from a competitor's website that would otherwise be cost-prohibitive. For example, a researcher could program a "bot" to collect data from a competitor's website by setting up the bot to solicit the online web store and pre-order every single permutation of products to identify pricing information as well as pricing bundling strategies. Another method would be to set the "scraper" to collect all data from the competitor's website by copying its HTML code.

The use of bots generates a range of ethical and legal issues. From an ethical perspective, it is frowned upon to use and solicit a competitor's resources in a manner exceeding that of an "average consumer." As such, you might contact customer services to ask a few questions, but you couldn't automate a "bot" to generate hundreds of queries on a website to gather the information you want.

There are also legal considerations: using a bot will most assuredly go against a website's terms of service, which explicitly prohibit the use of this type of tool. Courts in many U.S. states have interpreted the terms "without authorization" and "exceed authorized access" in a rather broad fashion, leaving the researcher exposed to lawsuits and legal penalties. While enforcement varies from one jurisdiction to the next, the researcher should take into account that he has crossed into a very gray area of market research.

3.3 Internal Secondary Data

Internal secondary data exists and is stored inside the organization. It is available exclusively to the organization, and is usually generated and collected during normal business activities. Internal sources of data should always be investigated first because they are usually the quickest, most inexpensive, and most convenient source of information available.

Sources of internal secondary data include

■ *Sales data*: If the organization is commercializing its product, it has access to an invaluable resource, its own sales data. This data is already collected by the organization and organized in a way that is useful and extractable by a market researcher. Some of the data that the researcher can look at includes sales invoices, sales enquiries, quotations, returns, and salesforce business development sheets. From this information, territory trends, customer type, pricing and elasticity, packaging, and bundling impact can be inferred. This data can be useful to identify the most profitable customer groups and which ones to target in the future.
■ *Financial data*: All functioning organizations have accounting and financial data. This can include the cost of producing, storing, and distributing its products. It can also include data on research and development costs and burn rate.

◼ *Internal expertise*: Mid-sized organizations will often have inside expertise, that is, personnel who have been with the organization for a while. These people can be tapped and interviewed to get more information on past initiatives, products, lost customers, or any other topics. Frequently referred to as the organizational memory, they often have a wealth of undocumented knowledge that can be harvested. They might be aware of internally produced reports that might be of use, past projects, or failed product initiatives.

One of the weak points of internal data is inaccuracy, due to the fact it may be dated or the way it was collected. Also, while most of the time data can be ported from internal systems to market research data analysis tools, some legacy systems may make data conversion especially challenging. Finally, there are confidentiality issues: some companies employing third-party researchers may hesitate to "open the books," and share either consolidated data or limited data sets.

3.4 Evaluating Your Secondary Research

Most secondary research, by its nature, is generated by a third party (the exception being the internal secondary data we discussed in Section 3.3). As such, it is important to verify its accuracy before using it. For this, we propose the use of data triangulation.

Data triangulation is the validation of data through cross-verification of two or more independent sources to ensure its accuracy. For example, to estimate market growth, a company could use multiple market reports, consolidating them with key opinion leaders' market growth estimates and information gathered from government websites. Data triangulation is very useful to strengthen the company's assumptions going forward.

> **On the importance of triangulating your data ...**
>
> Triangulation is an important factor when we look at market estimates. We often get presented data with a lack of triangulation, and this definitely raises issues. If a market assessment is evaluated through multiple vectors, we definitely see some value in it.
>
> *Ajit Singh, Managing Director and General Partner,*
> *Artiman Ventures.*

Data triangulation is especially important when dealing with sources of information that are akin to rumors and hearsay: information found on blogs, social networks, and discussion groups should never be taken at face value, but rather used as leads for further research. It has become so easy to manipulate these types of media (either purposefully, or though lack of rigorous fact-checking), mainly because of the way social media generates revenue—more clicks on the web pages means more views, which means more revenue from ads. Hence, popularity is a much more important metric than the truth. The information found on social media should never be implicitly trusted.

Of note, it can also be interesting to triangulate primary and secondary data as a way to verify data accuracy. One way to do this would be to collect some key pieces of data, and discuss them with key opinion leaders during an interview to get their opinion and perspective.

Finally, when triangulating data, it is crucially important to make sure that the sources used to triangulate do not refer back to the same initial source. Using a series of news articles and blog posts that all refer back to the same initial market report would not enable triangulation. Likewise, more and more media stories are found to be inaccurate, having implicitly trusted blogs and built stories on inaccurate information.

To illustrate the importance of triangulation, a few years ago, I was updating a total available market/serviceable available market (TAM/SAM) model for a client. He had pre-pared his own model a number of years earlier, and was looking to update it as well as to perform an independent validation. But we had an issue as my current estimates were much smaller than his previous estimates. As he believed his market was growing, he couldn't understand why my esti-mates were so much smaller. After some research, I found that one of the market estimates that he had used was based on a study that defined his market segment much more broadly than what he was targeting. As such, as my client had used a single point of reference as his starting point, his model grossly overestimated his market.

3.5 Limits of Secondary Research

As mentioned in the first few pages of this chapter, secondary data has a number of issues that have to be considered. There are issues related to availability, accuracy, trust, and bias.

First, there are issues of availability: not every topic will have secondary research data that exists and fits the needs of the researchers. The more specialized the topic, the more difficult it is to gather data. As such, careful consideration has to be given to how a topic has been defined. For example, if you are researching the generic manufacturing industry, and a report describes the market size, there could be issues with how the market was defined (does the author include bio-similars? Does the market study make a distinction between patented drugs and over-the-counter [OTC] drugs?). The researcher has to double-check the definitions used in the sec-ondary data he chooses to use.

Second, there are issues of accuracy, or the lack of data verification. A survey that you generate by yourself is verifi-able, as you are able to revise your numbers quite easily. And you can trust the data that you generate as you were the one who gathered it. But a secondary market report or a news article cannot be verified easily. You do not have access to the building blocks, and have to rely on the publisher's honesty and accuracy. If the author is biased, has been manipulated, or has made an error, the market researcher will be at the mercy of the data that he is using. Media references, especially those that are very close to social networks, should be used as information leads.

Thirdly, secondary data may not be sufficiently detailed or have such a small sample size as to not be useful. Like checking a scientific article, checking the sample size on which the author is basing his conclusions is critical when evaluating data. A data reference point without sufficient background on methodology should never be used.

Finally, as the author of market research must be careful to keep his bias in check, it is always useful to check for any potential bias in the data source being used. Some organizations and authors have their own agenda, which would taint the quality of information being shared.

Chapter 4

Analyzing Data

Data analysis is the process of evaluating, transforming, classifying, and modeling data. The current chapter will specifically deal with the evaluation to classification process, while presentation models will be discussed in depth in Chapter 5.

Before diving into analysis, it's important to note that the market research process can be iterative. It is possible to "jump" back and forth between the collection and analysis of data. For example, during initial data analysis, it is possible to find out that some new research areas or some of your data is insufficient, and a new data collection effort is necessary. The researcher should account for this possibility.

Case in point, a while back, a client had done a web survey to identify some trends related to salary and incentives in healthcare. He had collected an impressive sample (over 3000 entries), and I was brought in to analyze the data. Nonetheless, during the initial analysis of the data sample, the demographic data revealed that the overall weight of U.S. East Coast respondents was too large compared with the other U.S. geographic regions. As such, my client had to proceed with a second data collection effort, this time focusing on the other regions, if he wanted to obtain a representative sample.

In this chapter, we will provide an overview of data analysis. Our objective is not to turn the reader into a statistical analyst guru, but rather to give him a better sense of how to read the data he has collected, and some tools to better understand it. As such, we will be going over the basic elements of data clean up, followed by some words on quantitative and qualitative data analysis, and closing with some barriers to effective data analysis.

On the importance of your own data analysis ...

We're impressed with what some entrepreneurs are able to find using a limited amount of resources. But the value of their research efforts goes beyond what they show us in terms of market data and analysis. It demonstrates that they understand their environment, attests to their methodological approach, and that's a great asset since it is indicative of a strategic mindset and a curiosity to always look for the information they need.

Nicola Urbani, Investment Director at Gestion Emerillon Capital Inc.

4.1 Initial Data Analysis

Once your data collection is completed, it is quite tempting to immediately start data analysis. After all, the objective of the project is to answer one or more market research questions, and once you have your data in hand, it feels as if the answers are a footstep away.

But taking time to clean up your data is crucial, and this has to be your first step before data analysis. We call this step the *initial data analysis* phase. While often undocumented in the final report, some researchers state that as much as 80% of the time allocated to the statistical analysis process is actually spent on data cleaning and preparation.* As such, it

---

* Huebner, M. et al. 2015. A systematic approach to initial data analysis is good research practice. *The Journal of Thoracic and Cardiovascular Surgery*, 151(1): 25–27. http://dx.doi.org/10.1016/j.jtcvs.2015.09.085 (Accessed December 21, 2016).

is essential that your research plan includes a budget (in both time and money) for this initial data analysis.

Note that at this stage, the researcher is not analyzing the data itself to find the answers to his research questions, but rather focusing on the data set to evaluate its validity and adjusting it as needed (while documenting any adjustments that are made). This is a crucial step, especially if more complex data analysis will be done afterwards.

There are two main steps to initial data analysis: (1) cleaning up the data (which itself is split into a screening phase and a diagnosis/editing phase) followed by (2) the preparation and formatting of data for analysis.

4.1.1 Cleaning Up the Data

Cleaning up data consists of assessing the data to check for errors and inconsistencies, and then developing solutions to address the errors as well as documenting how you will be handling them.

Before starting to clean up your data, make sure you have backed up the original data set: you will most likely generate multiple different versions of your data throughout the data analysis effort, but you have to make sure that you always have a copy of the original data. That way, if you need to trace back some original data that has been transformed, you can access the original data to do so.

After backing up the data, the next step of cleaning data is to ensure that it is error-free. This is called the screening phase.* You have to identify any irregular patterns or inconsistencies, and you have to resolve these issues before moving forward. For example, you might notice that one of the data columns is empty, which was due to an error in importing the

---

* van den Broeck J, Cunningham SA, Eeckels R, Herbst K. 2005. Data cleaning: Detecting, diagnosing, and editing data abnormalities. *PLoS Medicine*; 2(10):e267. http://dx.doi.org/10.1371/journal.pmed.0020267 (Accessed December 21, 2016).

data from the online survey software to the data analysis software. Other common errors include different variable types in the same columns (mixing numeric and alpha numeric are the most common mistakes) and having a shifted column due to a conversion error.

Once you are satisfied that the database is error-free, you can start familiarizing yourself with the data, identifying more specific issues and correcting them, as well as removing incorrect entries: as a rule of thumb, no more than 5% of the total number of data samples should be rejected.*

This is the diagnosis and editing phase of the data and includes the following components:

1. Make sure data from multiple data sources is integrated in a consistent pattern; for example, if there were multiple people doing interviews, you have to make sure the data is entered the same way for each interviewer.

2. If, before data collection, the market research plan was to remove incomplete surveys, they should be removed at this time.† If no decision was made, look at your minimum sample size; if counting these incomplete survey responses is enabling you to reach your minimum survey size, consider whether you want to (and can) do a new data collection effort. If it is not possible (e.g., due to time constraints), any incomplete responses included should be cleaned up to avoid any compilation issues.

---

* Manikandan S. 2010. Preparing to analyse data. *Journal of Pharmacology & Pharmacotherapeutics.* 1(1):64–65. doi:10.4103/0976-500X.64540. https://www.ncbi.nlm.nih.gov/pmc/articles/PMC3142762/ (Accessed December 21, 2016).

† While not obligatory, it is more professional if your sample size stays the same across different tables, rather than having different sample sizes when going from one table to the next.

3. Check for inconsistencies in the data, such as duplicate records or multiple answers for the same Internet protocol (IP) address.
4. Apply any consistency checks that were built into the survey, removing inconsistent entries (see Section 2.3.7 for more information on building consistency checks).
5. Exclude any unneeded data, using the predetermined inclusion or exclusion criteria: for example, if the study was designed to generate a precise image of the U.S. market, remove any non-U.S. responses that you have collected.
6. Identify the validity of outliers in the quantitative data set. These might invalidate statistical models or create inconsistent results during analysis, so it is important to verify outliers individually. Furthermore, how these outlying values were identified and handled needs to be documented.
7. Scan qualitative answers to your survey to remove entries with unusable data. For example, some participants will answer "awesome" to every single qualitative question, while others will take a moment to complain about the company sponsoring the survey, or the survey itself. As a rule of thumb, I expect that anywhere from 2% to 3% of survey responses will need to be removed due to unusable answers.
8. Double-check group labels and consistency in naming for category variables; while this might sound trivial, it is a time-saver to take the time to format and correctly name data labels, especially during the data analysis phase.

Once these steps are done, you should check if the data sample size is lacking, sufficient, or excessive: it is possible that after cleanup, the sample size will no longer be the necessary size to reach statistical significance and a new phase of data collection will be needed.

4.1.2 Preparation of Data

The final step of cleaning data is the preparation of the data for more advanced analysis. While producing summary tables can be useful, the researcher might decide to adjust the variables for advanced analysis. Some of the data can be used in its unprocessed form, but it might be useful to rescale or standardize categories, transforming some data into a single variable, combining them into summary scores, or using more complex functions such as ratios.* As such, you might need to create new variables that will specifically address your research topic (by creating indices from scales, or combining data sets that are too small to be useful categories) and format the variables so they can be processed correctly by the data analysis software. This can include setting a value for missing data codes so that your software can handle them and correctly formatting data variables (so you don't have any numeric and alphanumeric mix-ups).

Demographic data is often the subject of aggregation, specifically variables related to geographic location, age, and educational level. For example, during data collection, a researcher might have collected the state and city of each respondent, but might later realize that this level of granularity is not useful for his research objective, and decide instead to recode the data according to major geographic regions. It is also possible to recode data if a subgroup is too small to generate useful information. For example, if you did a web survey of 3000 participants, and less than 1% of participants were aged 65 and older, it might be useful to aggregate them into the closest category and redefine it: so you could combine the data from the 55–64 years category with the 65 and more into a single "55 and more" category.

---

* Vach W. 2013. Transformation of covariates. In *Regression Models as a Tool in Medical Research*. Taylor & Francis Group: Boca Raton, FL, USA; 264–273.

4.1.3 Specific Issues Relating to Cleaning Up Qualitative Data

Qualitative data analysis is more complex than analyzing quantitative data. It is often produced on different medias (written notes, sound or video recordings, pictures), which first require transcription into electronic media.

The transcription should be done word for word, not just summaries as summaries imply the initial transformation of data: if the transcriber is summarizing data, he is introducing some bias by choosing what information to include, which words to summarize, and so on. Some other things to consider:

- The transcription document should include nominative information, such as who performed the interview, who was interviewed, as well as the time, date, and location the interview took place.
- During transcription, non-verbal cues that were noted by the interviewer should be included in the transcript, as these could prove informative later on: mark them distinctively so you can quickly identify them.
- If possible, try to transcribe transcripts yourself (or have someone who attended the interview or focus group transcribe the information), so you can add any thoughts or ideas that emerge during transcription.
- Even if you are trying to be as faithful to the original text as possible, you can edit out verbal tics such as "you know" or "it's like," as well as fillers (such as "uh," "hmm," and more) as these can prove quite distracting during analysis.

4.2 Data Analysis: Quantitative

Quantitative data refers to data that can be measured and numbered. Counting the number of potential clients for a product, calculating the number of products or doses a consumer uses each day, or the average distance a patient is willing to travel to visit a specialized clinic are all different types of quantitative data.

Quantitative data is easier to share since the results are easy to understand and easy to explain. There is a perception that quantitative data is "hard data" (vs. qualitative data that is perceived as "soft data"). Hence, market research often has a large quantitative component to convince the target public.

Explaining how to analyze quantitative data is an ambitious endeavor, and we cannot give a complete overview of the process in the limited space we have allocated to it here. There are many fine books dedicated to this topic exclusively, and any reader wishing to learn more could take a look at *Quantitative Analysis for Management* (12th Edition) by Barry Render (Pearson Publishing) or *The Quantitative Analysis of Social Representations* by Alain Clemence (Taylor & Francis).

This section starts with a presentation on the two different types of analysis: descriptive and inferential analysis. This is followed by some presentations on univariate and bivariate data analysis models, and some simple tests you can use to verify the significance of the results, such as correlation and regression models, as well as some interesting models that can be used to analyze your data. We conclude this section with a description of computer software packages that are available to analyze quantitative data.

4.2.1 Overview of Descriptive and Inferential Analysis

There are two types of quantitative analysis: descriptive and inferential analysis. When doing descriptive analysis, the information describes or summarizes the data that was collected, whereas inferential statistics are used to make inferences from the sample data for a whole population.

Descriptive analysis is used when the researcher is analyzing his data so he can identify patterns in the population he is studying. Contrary to inferential analysis, descriptive analysis will summarize the data in the sample itself, rather than inferring that it applies to a whole population. The advantages of doing descriptive analysis are multiple: first, doing descriptive analysis makes vast amounts of data both more manageable and organized. It is also straightforward, can be used to further research ideas, and lays the groundwork for more complex analysis.

In this book, we will be looking at descriptive analysis from a univariate analysis angle (with a focus on frequencies of variables, central tendencies, and dispersion) and a multivariate analysis angle (with emphasis on the relationships between variables).

The restriction of descriptive analysis is that it is limited to the data you are handling. It is used to report and describe data that you collected, but cannot be used beyond the current data. The researcher cannot use the data to interpret phenomena beyond the population from which the sample was taken. For example, if you tested how a product is used by a sample population, you could not consider the findings representative of the population at large using exclusively descriptive analysis.

Inferential statistics would be needed. So, while descriptive statistics are used to describe what is happening in the sample population alone, inferential statistics are used to infer what the population beyond the data sample is thinking. Hence, inferential statistics are techniques we use with our data set to make generalizations about the populations from which the samples were collected.

The limitations are that inferential analysis is uncertainty. While calculating the proportion of a sample in a specific situation is absolute, projecting it onto a population means that it becomes an estimate: building this estimate requires the researcher to make educated guesses, and this uncertainty will increase the likelihood of errors.

There are a number of techniques that market researchers use to examine the relationships between variables, thereby creating inferential statistics. We will look at some of the most valuable ones in market research: correlation, regression analyses, and general linear models.

4.2.2 Univariate Analysis

Univariate analysis is the simplest way to analyze data. All that it entails is the analysis of a single variable, which can be categorical or numerical. Since it is not being compared to one or more other variables, it is very descriptive in nature. Univariate analysis cannot be used to explain a phenomenon or to contextualize a variable in relation to other variables. Once data is compiled, it can be shared in a simple table form, as well as in visual representations such as bar charts, histograms, frequency polygons, or pie charts.

The most frequent univariate analysis methods are frequencies of variables, central tendencies, and dispersion.

4.2.2.1 Frequency Distribution, Central Tendency, and Dispersion

Frequency distribution is a simple data analysis technique that allows the researcher to get a quick impression of the data he has collected. Using frequency distribution, he can see how often the specific values are observed and their percentages compared to the overall sample. As such, frequency distribution enables the researcher to summarize each variable independently, and also makes it easier to subsequently engage in more complex analysis.

To illustrate frequency distributions, we will use data from a project I did a few years back. It was for a company in the dental health space. This client had developed an innovative dental product, and was looking to evaluate how much interest dentists would have in the product, according to various distribution models. Over 100 dentists (current and past clients) participated in an online survey. The survey included both quantitative and qualitative questions.

To validate interest in the innovation, the client had asked the following question: "How likely would you be to use [product description] if you had easy access to it and the turnaround times described above? (i.e. less than 90 minutes)" The data was compiled and illustrated in both a table (which is more detailed and descriptive) and a graph (a more visual format to share data) (Table 4.1; Figure 4.1).

Table 4.1   Frequency Distribution Example of Survey Data in a Classic Table

Research question: "How likely would you be to use [product description] if you had easy access to it and the turnaround times described above? (i.e. less than 90 minutes)"				
	Frequency	*Relative %*	*Valid %*	*Cumulative %*
Very likely	32	28.8	29.9	29.9
Likely	26	23.4	24.3	54.2
Somewhat likely	34	30.6	31.8	86.0
Somewhat unlikely	6	5.4	5.6	91.6
Not likely	9	8.1	8.4	100.0
Total	107	96.4	100.0	
No answers	4	3.6		
Total participants	111	100.0		

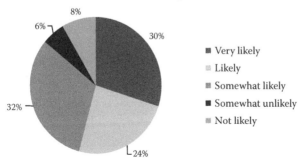

Current interest for product

8%
6%
30%
32%
24%

- Very likely
- Likely
- Somewhat likely
- Somewhat unlikely
- Not likely

Figure 4.1   Frequency distribution example of survey data in a classic pie chart.

Based on this univariate analysis, there seemed to be some strong interest in the current client base, as 86.0% of participants were either "somewhat likely," "likely," or "very likely" to use the product developed by my client.

4.2.2.2 Central Tendency

There are many ways to find the central tendency of a data sample. You can use either the mean (the sum of the values divided by the number of effective values), the mode (the most frequently occurring value), or the median (the value separating the upper half of a data sample). This is the traditional method of presenting scoring data and giving a sense of the general appreciation in a population (e.g., statements like: "participants in our survey gave an average of 7 out of 10 for the overall look of our new product label").

Using a mean is the most popular way to calculate central tendency, as it includes all the values of your data sets, but it is particularly sensitive to the presence of outliers. The outlier can skew the mean, rendering the data unusable. Another reason for not using the mean is if your data itself is skewed: if, for example, half your sample would be willing to pay $10 for your product, and the other half would be willing to pay $20, a mean of $15 would not reveal that you probably have two distinct market segments in your data rather than a single monolithic segment.

As for modes, these demonstrate the most popular option in the sample. The problem when using the mode is when it is not unique: if you have two data sets with the highest frequency, it can be challenging to interpret the data. Another issue with the mode is that it does not give a good read of the central tendency: if you asked each participant to give a score of 1–10 on the quickness of a product, and then half responded 1 out of 10, the mode would not be representative of the central tendency.

The median is not affected by the presence of outliers, so it gives a better idea of a typical value. It is also better to use when the distribution of the sample is skewed. The main issues when using the median (rather than the mean) are that it is less popular than using the mean to communicate the central tendency, so people might not be familiar with its meaning. Also, a medium is much more complex to calculate if you are not using software, which honestly should not happen too often if you are preparing a market report.

4.2.2.3 Dispersion

Dispersion is the calculation of the distribution of values around a central value. The three main dispersions that you can calculate are range (the distance separating the highest and lowest value), the variance (the average of the squared differences from the mean), and the standard variation (which is the square root of the variance and measures how much the numbers are spread out).

Calculating dispersion is important to understand the spread around the mean, or the variation around central data. The mean of a data set indicates the average, but the dispersion will supply you with invaluable information on how much of your data sample is grouped within a certain distance of your mean. The lower your dispersion (Figure 4.2), the more your mean is representative of your central tendency. The higher the dispersion (Figure 4.3), the more dispersed your data set is.

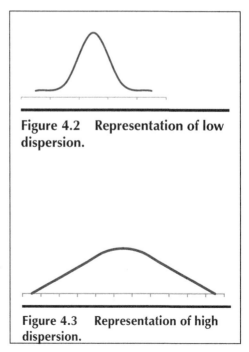

Figure 4.2   Representation of low dispersion.

Figure 4.3   Representation of high dispersion.

As an example, a project I worked on surveyed over 5000 individuals on their Internet usage. One of the questions asked to participants was to estimate time spent online. Across all age groups, the average was 19 hours, with a standard variation of 12 hours. This meant that most of the data sample had answered between 7 and 31 hours spent online, giving us a rough idea of the time spent online for the surveyed population.

4.2.3 Multivariate Analysis

Multivariate analysis is the concurrent analysis of two or more data sets. While univariate data analysis is used to *describe* a phenomenon, multivariate analysis is used to *explain* the phenomenon.

The objective is to compare two (or more groups), and to identify relationships between the multiple data sets. The analysis exposes how different subgroups respond to some query. The simplest method of doing multivariate analysis is cross-tabulation (or contingency tables), while more complex tools include correlation and regression analysis.

4.2.3.1 Contingency Tables

A simple way to do multivariate analysis is to build contingency tables, which are multidimensional frequency distributions in which the frequencies of two (or more) variables are cross-tabulated. These simple tables can help the researcher find the relations among different variables. Contingency tables are called pivot tables in Excel. A useful guide to creating a pivot table to analyze worksheet data can be found directly on the support.office.com website. For ease of use, you can follow the abbreviated link: goo.gl/XykXIv.

To properly illustrate a contingency table, we are sharing a table illustrating bivariate analysis (Table 4.2).* The data

---

* Bivariate analysis refers to the analysis of two variables. If three or more variables are analyzed, then it is called multivariate analysis.

for this table was for a project I did a few years back, the theme of which focused on "what women want." Surveying Australian mothers, one of the topics the survey dealt with was identifying how mothers got information relating to their pregnancy. This specific table cross-tabulated data from two questions:

a. Did you use this information source to get more information relating to your pregnancy? (Check all that apply.)
b. What is your age?

At a glance, there are four apparent key learnings we get by doing a cross-tabulation (Table 4.2):

a. It seems that women younger than 39 years old are much more likely to use online resources to look for information than their older counterparts.
b. It seems that women younger than 39 years old are much more likely to consult their mothers for advice than their older counterparts.
c. Only half of the participants contacted the Department of Health to get information relating to their pregnancy.
d. There is an overwhelming use of pregnancy books and manuals across all age groups.

Overall, a contingency table is an effective and simple way to find underlying information that univariate analysis would not have been able to discern by using the data from two distinct data sets.

4.2.3.2 Correlation

Correlations are an extension of a cross-tabulation analysis. While building a contingency table might uncover a potential relationship between multiple data sets, correlation analysis is necessary to confirm the relationship between the variables.

Table 4.2   Example of a Bivariate Contingency Table: How Women Find Information Relating to Their Pregnancy versus Age of the Respondent

		Age		
		18–24	*25–39*	*Over 40*
Criteria	Used			
General online resources	Yes	81	800	72
	No	38	477	151
Department of Health	Yes	40	421	59
	No	79	856	164
My doctor	Yes	79	838	144
	No	40	439	79
My friends	Yes	71	797	142
	No	48	480	81
My mother	Yes	86	762	92
	No	33	515	134
Other family members	Yes	56	541	77
	No	63	736	146
Pregnancy books and manuals	Yes	88	1120	196
	No	31	157	27
Pregnancy-focused websites and e-mail newsletters	Yes	87	896	94
	No	32	381	129
Pregnancy forums and communities	Yes	62	504	39
	No	57	773	184

To confirm the existence of a correlation between two variables (as well as regression and general linear relationships), the use of specialized software such as SPSS or JMP is recommended, although it is possible to use Excel as well. It is also possible to use specialized websites, such as the social science statistics calculator found at www.socscistatistics.com/tests/pearson/.

When calculating correlation, for each data set, the software will provide an answer ranging between 1 and −1. The closer the answer provided is to 1, the stronger the relationship between the two variables. The closer it is to zero, the more it means that there is no relationship between the two variables. If you obtain a correlation of −1, this means that your two data sets are negatively correlated, meaning the higher that one value goes, the lower the second one goes. As a rule of thumb, if you obtain a correlation value of 0.5 or higher, this means that there is a significant relationship. This becomes interesting as you start to calculate the strength between multiple sets, and you can statistically see which relationships are stronger and determine key elements, or those that are important for the participants in your research.

There are two major caveats to calculating correlation. First, *correlation is not causation*. This means that you might have found that two data sets have a relationship, but there is no proof that one element is causing the other. There could be multiple reasons why your data is correlated. Second, calculating correlation is only possible if your data is in a linear relationship. A good alternative if you believe that there is a relationship between two data sets, but that your calculations or correlations don't support it, is to do a scatter plot of your data so you might be able to visualize a non-linear relationship.

4.2.3.3 Regression Analysis

Regression analysis is another way to estimate the relationship between two variables. It is a test used to see how one independent variable affects the other dependent ones. Some prefer it to correlation since it statistically demonstrates the goodness of fit (adjusted R Square). Regression analysis can

Table 4.3   Guidelines for Interpreting Correlation Coefficients

If the coefficient is ...	Then the correlation is ...
Between 1 and 0.7 then	Strong
Between 0.7 and 0.4 then	Moderate
Between 0.4 and 0.2 then	Weak
Less than 0.2	No correlation

also be used for forecasting, using the regression equation as a simple linear model.

One of the main differences between correlation analysis and a regression analysis is that while correlation analysis quantifies the relationship between variables, a regression analysis can be used to predict the relationship between variables, especially if one of the two variables is one that you manipulate (e.g., time or price).

As a rule of thumb, the guidelines for the interpretation of your regression factors are as shown in Table 4.3.

4.2.3.4 General Linear Model

General linear models are used to assess the effect of predictors on one or more continuous variables. They include tests such as the *t*-test (which is used to examine the difference between the means of two independent groups) and the ANOVA test (tests the significance of differences between the means of two or more groups). These are useful to assess the effect of several predictors on one or more continuous dependent variables.

4.2.4 Software for Quantitative Analysis

There are two types of software that you can use for analyzing quantitative data.

First, it is possible to do a lot of basic and even some intricate analyses using spreadsheet software such as Excel.

Spreadsheets can calculate anything from simple averages (with the AVG and MED functions) and frequencies (using the FREQUENCY function) to more complex calculations such as standard deviations (STDEV), correlations (using the CORREL function), or even trends (using the TREND function, calculating regressions). There are a number of resources online on how to use advanced functions, but one of the more interesting books I have used is *Excel Statistics: A Quick Guide*, written by Neil Salkind and published by Sage Publications.

The main advantage of using a program like Excel is that many users are already familiar with spreadsheets, so the learning curve to using it is smooth. Also, integrating data from other software (such as online websites used to collect survey data) is pretty straightforward. Finally, as you are already formatting your data in your spreadsheet, preparing your data for presentation models (tables or graphs) is streamlined. As such, spreadsheets are a good alternative for companies that do not engage in statistical analysis on a regular basis.

For more complex analysis, or if you engage in quantitative analysis on a regular basis, obtaining a statistics software package is a necessity. There are a number of them available, but here are a few of the more popular ones.

- One of the market leaders is IBM SPSS Statistics (www. spss.com). It is a powerful statistics software package used for statistical analysis in academic and business environments. It is quite popular with scientists and marketers alike, and contains a full suite of tools that are useful for any researcher. It has a wide range of charts and graphs to choose from and efficient access to statistical tests. While powerful, it is one of the costlier solutions out there.
- SAS (www.sas.com) is another prevalent analytical software program that is used for advanced analytics and analysis. It is positioned as one of the most powerful

software programs available, but is quite complex to program and utilize. It is one of the hardest software programs to learn, but offers great data management and the ability to work with multiple files simultaneously.

▪ A less expensive option, albeit still powerful, is JMP (www.jmp.com). Like SPSS, JMP has a number of options for data analysis and produces a full range of charts and graphs. It is very straightforward to use and somewhat less expensive, and easily imports data from most formats, including SPSS tables.

There are also a number of free statistics software packages available, but be aware that the learning curve is steeper than some of the commercial packages available. One of the popular free alternatives is PSPP (http://www.gnu.org/software/pspp/pspp.html), which, while not as powerful as the commercial alternatives, does have a similar look'n'feel to SPSS, which makes the transition easier. It also has an easy point'n click interface.

4.3 Qualitative Data Analysis

Qualitative data is information that is subjective and subject to interpretation. It can include anything from stories to words, observations, pictures, and even more peculiar sources such as songs, stories, and poems. Examples of information you can collect through qualitative analysis include personal preferences in consumer purchases, the impact of quality on customer purchasing decisions, or the influence of packaging color on acquisition decisions. Data collected through interviews, focus groups, Delphi groups, and observation is usually qualitative in nature, but as we will see, it is possible to codify the collected data so it can be analyzed using quantitative data tools.

Qualitative research is particularly useful when the objective is to find out what people are really thinking, identifying future trends, and speculating on competitive threats. Strategic thinking and careful analysis is needed to identify those trends. In the next few pages, we will go over a proposed framework for qualitative data analysis, followed by a discussion of computational tools available to enhance qualitative data analysis.

4.3.1 Qualitative Data Analysis Process

Quantitative data has the advantage of being simple to analyze. It is often collected in a format that comes pre-coded and ready to analyze, whereas qualitative data needs to first be codified and labeled before proceeding with the analysis. To be able to code qualitative data, the researcher will need to build a classification framework, following a process to categorize verbal or behavioral data to enable the classification, summarization, and tabulation of his data.

The four steps of the basic qualitative data analysis process are (1) familiarization, (2) identifying a framework, (3) sorting the data into the framework, and (4) using the framework to complete the descriptive analysis.

Some researchers suggest doing an early analysis step before diving into the complete qualitative data analysis. This implies analyzing some qualitative data during data collection. They believe that it is an effective method to optimize ongoing research and generate new ideas for collecting better data. It can also give you an early idea of what your analysis framework will look like. I believe this should be avoided in studies that are exploratory in nature, to ensure that the first few interviews do not direct you down a path that hasn't yet been validated. For example, if you do an interview with a few doctors to explore the important factors for signing up for your new service, and they focus on the login and payment screen, there is no indication that these concerns are valid throughout your population, and focusing on them too early in the study might bias the final outcome.

4.3.1.1 Step One: Familiarization

The first step of qualitative data analysis is to familiarize yourself with your data and what it looks like, and then start to visualize the necessary effort for data analysis. This means re-listening to multiple interviews, reading multiple interview transcripts, and reviewing open-ended survey answers as well as rereading the notes you collected during the research effort. If you worked as part of a team, a debriefing meeting will be useful to discuss high-level impressions as well, but be careful that these discussions don't bias the next analysis steps.

4.3.1.2 Step Two: Identifying a Framework

The qualitative data framework is the coding plan that you will be using to organize your information. During this step, you will be identifying high-level data patterns that you will use to build your first framework.

This framework will consist of *codes*, which are tags or labels that are used to assign meaning to qualitative information. They are usually attached to pieces of text, such as words, expressions, phrases, and even paragraphs. They can take the form of a simple word, or can be more complex, in the form of expressions and metaphors. These codes are used to create meaning through clustering: the more often a code appears during analysis, the more meaningful the meaning behind it. These code clusters are then used to analyze data, identify patterns, and generate recommendations or concepts.

When developing your coding, it is possible to use three types of codes:

1. Description codes: Codes that describe the phenomena, and require very little interpretation.
2. Interpretative codes: When you have a bit more knowledge about what you are researching, you can start assigning more complex coding by using codes that integrate some interpretative elements.

3. Pattern codes: These are codes that you can start using when some of the more complex elements such as patterns, themes, and causal links are identified in more depth. This might require you to go over some earlier analysis to verify if some codes need to be updated.

There are multiple methodologies that can be used to build a framework and start identifying your codes, but we will be going over three of the more common ones. These are manual iteration, automated text analytics, and directed content analysis.

Method 1: Manual iteration

Manual iteration is the most complex and lengthy method to develop a framework, but it is the one that will less likely result in you having to go over your data twice.

The first step is to read through a few records to get a high-level appreciation of which phenomena are occurring most often. If the bulk of your qualitative data is interviews, review a few transcripts and start identifying key trends. If it's focus groups, read through the transcripts and moderator notes to identify those common elements. If it's a survey with qualitative open-ended questions, read the first 100 responses, and identify keywords, ideas, and thoughts in a separate column of the data analysis tool that you are using. Afterwards, go through the new column of keywords that you generated to identify patterns and build a framework for analysis. If a consistent framework cannot be found, read a few more interviews, or a second group of 100 survey responses, and then look through your notes to identify that pattern.

To illustrate the process in an online survey, I have included a short extract from a project I analyzed a few years back.

To give some context, my client, a company active in event organization and media platforms targeting the Generation Y public, had organized a survey targeting its subscribers. One of the issues they investigated was the purchasing patterns for healthy drinks. Hence, they had asked participants to think about the last drink they had purchased, and the reason that had led to that purchase over another competing brand or product. My client had collated the data, and brought me in to analyze it.

Hence, the first step was to build an analysis framework. I chose to do it through manual iteration since the sample was quite large and complex, and the context was exploratory. To demonstrate, I have included a few statements taken from the survey, followed by some tags that I felt were suitably related (Table 4.4).

Table 4.4   Example of Building a Codification Framework through Manual Iteration

Statement	Keywords
A friend recommended it to me, next time I saw it in a shop I tried it, and he was totally right.	Friend's recommendation, convenience
Because they were the cheapest!	Pricing
I bought it because of brand loyalty, design, hype, health.	Brand loyalty, design, popularity, health
I am trying to get healthy and I had heard about it through family.	Health, family recommendation
Word of mouth from friends that the product is superior to others	Word of mouth, friend, recommendation
Recommended reviews online, a well-known/trusted brand around for a long time, good price.	Online review, brand reputation, trust, price
I bought it because it was on sale and mostly because I don't want to fill my body with industrial waste anymore.	Price, sale, health, ethical

Other patterns were identified around themes such as "locally made," "great website," "celebrity endorsement," "advertising," and more. Not all of these keywords were kept in the final version of the framework: some keywords, which were present in the first few entries, had very little presence in the survey responses overall, and were discarded. Hence, since the process is iterative, some keywords that are identified in the first step might be dropped later on, while others are added during the more in-depth analysis. Some overlap might also occur, as well as splitting up some keywords of your framework.

You might note that there is a high level of subjectivity in this process. While I use "friend's recommendation" to denote the phenomenon of "a purchase due to a recommendation of a non-family-related person," you might prefer using the code "third-party recommendation" to widen the definition to "a purchase due to a recommendation of a third party." This is all part of the strength and weakness of qualitative data. Future iterations will enable you to refine your framework—you might find that there is enough of a distinction between friends and parents to warrant a second category. Or you may not. Flexibility is the key at this stage.

This is also the reason why two researchers need to work in tandem on some of the bigger projects. By going over the work independently, and then combining the results afterwards, the researchers are able to develop some level of consistency.

The framework that you build will allow quicker analysis of the responses that you have collected. By the end of this step, you will have identified a series of keywords, and will be ready to advance to step three.

Method 2: Automated text analytics

This method can be used with the qualitative analysis of data sets for which the answers are short yet in high quantity, such as qualitative answers in an online survey. The idea is to use a word cloud generator.

Word cloud generators are platforms that count the number of times each word appears in a data sample, assign it a relative density, and then generate a visual representation according to the word density of the top words. There are many free word cloud generators that exist online such as WordItOut (worditout.com/word-cloud/create) and WordCloud (www.jasondavies.com/wordcloud/), which you can use to generate your own word cloud for both analysis and presentation purposes.

After choosing your word cloud generator, copy a random number of entries for the topic you are analyzing into the text box. The platform will generate a word cloud, giving more prominence to the words that appear more frequently. These words can be used as a basis for the first framework for data analysis. Be advised that while this method is much easier and faster than manual iteration, this shortcut will likely overlook some keywords, and some revision of your framework might be needed later on.

To illustrate, using the same data set as used in the first method we generated a word cloud using 100 random answers from participants. The example was generated using the WordCloud website. Remember to exclude common words to increase the visual effect of the word cloud. In this case, we excluded words such as "because," "around," and other commons articles that are generally not important (Figure 4.4).

Figure 4.4   Word clouds generated by online word cloud generator "WordCloud."

A quick review of the word cloud lets us appreciate the emphasis that participants place on the quality of the product, the brand name, the reputation, recommendations, and reviews, for example. Using the word cloud, we can start the next step of analysis with a half-dozen or more code words such as quality, price, reviews, friend's recommendation, brand, taste, trust, and healthy, for example.

Method 3: Directed content analysis

Finally, some researchers use directed content analysis. Using this methodology, some of the codes are developed before data analysis, using theory and a review of the literature/ secondary research to build the question guide and the theoretical framework to guide analysis. Using this method, additional codes are added as the analysis progresses. The advantage of this method is that the existing framework lets the researcher jump into the analysis phase faster, and the questions guide can be built according to the developed analysis framework. Nonetheless, this is less useful in exploratory contexts, and can bias some of the data collection as well as the analysis.

Coming back to our example presented earlier in this chapter, if we had decided to use directed content analysis for our project, there are several ways we could have built our framework:

■ If we had built it on the client's experience (hence a "hypothetical framework"), the client and I would most probably have used terms such as *cool, price,* and *friend's recommendation* as starting points for our framework. (Based on a quick review of our e-mail conversations, these are the terms he expected to emerge.)

■ Alternatively, I could have reviewed an article such as "Marketing to the Generations"* from the *Journal of Behavioral Studies in Business* and identified "prestige," "uniqueness," "pricing," and "referrals" as starting points.

In both cases, the results are the same: a pre-existing framework is imposed over qualitative data. Even if there is a speed advantage, there is a potential negative impact on the analysis of exploratory data that should not be underscored.

4.3.1.3 Step Three: Coding the Data Using the Framework

Once a suitable framework has been built, it is time to assign the codes to your data. As this is an iterative process, expect to modify your framework during analysis: you will most likely remove, merge, and add codes all the way during your analysis process. This might mean that some data elements have to be reviewed multiple times. This is a normal step of qualitative data analysis. In this section, we will illustrate two examples of coding qualitative data, one for interview data and one for survey data.

To code interview data, I usually use a three-column analysis worksheet. The first column has coding that is attributed to a sentence. The second column is the interview itself, with phrase segments allocated to phenomena, and the third column is general comments and thoughts that emerge during analysis. These thoughts are often very useful during the iterative process, to help adjust the coding categories, to help define them precisely, and to prepare for the next step, pattern coding.

---

* Williams K., Page R. 2011. Marketing to the generations. *Journal of Behavioral Studies in Business*, Volume 3–April, 2011. http://www.aabri.com/manu-scripts/10575.pdf (Accessed January 12, 2017).

To illustrate this process, here is a short extract analysis of an interview I did a few years ago. It was for a company planning to offer outsourcing services to large companies, and needed to better understand the trends impacting the need for specialized outsourcing abilities. The company needed to understand decision-making processes and identify growth opportunities. A review of the literature had enabled us to build the framework shown in Table 4.5 beforehand.

The codification then enables quick clustering and analysis: instead of looking through hundreds of pages of interviews, you can quickly scan and look for patterns or codes that reappear often, and then group these comments to craft your story. Codes make the interview more intelligible, and the recurrence of certain codes signals the emergence of regular themes (Table 4.6).

Survey analysis follows a similar approach. Following up on the survey on purchasing patterns for healthy drinks, we coded all the responses from the participants. Each statement was individually analyzed, and assigned a code from our framework. As the analysis proceeds, new codes can emerge, and new ones can be created from consolidation. For example, during the consolidation stage, the categories "user reviews" and "online reviews (blog)" was consolidated into user reviews, as the number of responses did not justify keeping those two categories separated.

Table 4.5   Sample Coding Framework

Code	Definition
REL	Relationships
OUT.GO	Outsourcing to global players
OUT.LO	Outsourcing to local players
VIR	Virtualization

Table 4.6 Example of Analyzing an Interview Extract

Codes	Text	Notes
	JFD: One of the key issues I wanted to explore is if there are any preferences for local providers versus international providers.	
OUT.GO OUT.LO VIR	MM: At our company, we prefer global providers to match our global operations, but then we prefer to use global providers that have a presence near our strategic offices.	N1: Looking for global players that are local, best of both worlds …
	JFD: Why is that?	
	MM: We have found that virtualization is a thing, but sometimes problem resolution requires a face to face conversation. Also, it's just better to be able to interact directly.	
	JFD: Has that shifted the nature of the outsourcing you do?	
REL	MM: It has. It means looking for the right partner. I believe organizations will shift strategy, taking a longer-term view of sourcing options instead of a one-time decision on which functions to outsource and those to keep in-house. Provider selection is changing. It will take greater effort to cut the field to four finalists who receive the request for proposal (RFP).	N2: Seems to be a recurring theme, focus on the long-term relationships.

To ease codification, we built an analysis matrix. Qualitative data was stacked on the lines, as columns were each assigned a framework item. As the process was iterative, it was possible to add codes as new elements were found. To best illustrate this, I have included a sample extract of the matrix that was generated (Table 4.7).

4.3.1.4 Step Four: Use the Framework for Descriptive Analysis

Once data is analyzed, the codes are used to create significance through the use of clustering. The more often that codes appear during analysis, the more expressive the concept behind them. These clusters are then used to identify patterns and generate recommendations.

For example, in Table 4.7, we see clusters emerge around the concepts of brand loyalty, pricing, and health. That is one of the advantages of using a matrix data analysis, as it provides a medium for an initial visual interpretation of data. The second advantage is that the data can then be converted into quantitative data. By compiling the tags, we are able to create a table, and generate the data as shown in Table 4.8.

As for interviews, the next step is pattern coding. Here, the objective is to group the categories into smaller sets of themes and constructs. Hence, during this step, recurring phenomena emerge, and you can see the most important ones. Finally, remember that it is very important to carefully define each theme; if they are not properly defined, some blurred lines might complicate the analysis, and render the framework invalid, or ill-defined.

4.3.2 Using Computer Software to Assist in Qualitative Analysis

Qualitative analysis software greatly enhances a researcher's ability to process and analyze large amounts of

Table 4.7 Example of a Matrix of Survey Data Analyzed

	Specifications and Features	User Reviews	Brand Loyalty	Personal Favorite/Wanted It	Great Quality	Family Recommendation	Local-Made/Local-Business	Pricing	Availability / Convenience	Looks Great	Friend Recommendation	Online Reviews (Blogs)	Healthy
A friend recommended it to me, next time I saw it in a shop, I tried it, and he was totally right			X						X		X		
Because they were the cheapest!								X					
I bought it because of brand loyalty, design, hype, and health	X		X							X			X
I am trying to get healthy and I had heard about it through family						X							X
Word of mouth from friends that the product is superior to others					X						X		
Recommended reviews online, a well-known/trusted brand around for a long time at a good price			X					X				X	
I bought it because it was on sale and mostly because I don't want to fill my body with industrial waste anymore								X					X

Table 4.8   Example of Compiled Qualitative Data: Top Ten Reasons Readers Purchased Their Last Drink over Another Competing Brand or Product

Reason	N	Percentage
Pricing	234	23.4
Brand reputation	166	16.6
Great quality	150	15.0
Friend recommendation	97	9.7
Specifications and features	88	8.8
I needed it	79	7.9
User reviews	78	7.8
Familiarity	77	7.7
Personal favorite/I wanted it	71	7.1
Advertising	61	6.1

data. It can ease the burden of manual transcription, and reduce miscalculations due to human error. Automated searches, auto-coding, and integrated analysis tools all facilitate researcher tasks.

The use of software for data analysis should be guided by many factors, such as the type and amount of data being manipulated, the time needed to learn to use the software versus the time analysis is expected to begin, cost constraints, and the need to share the data across multiple researchers. Nonetheless, one has to realize that while using qualitative software data will make the codifying and analysis process that much easier, it also distances the researcher from his data, and he loses some of the opportunity to understand his data.

Some of the most popular qualitative data software programs include

■ Atlas it (www.atlasit.com): Allows codification of data, as well the visualization of complex relationships between data sets.

■ QDA Miner (www.provalisresearch.com): Includes both coding and analysis components, and enables teams to share data virtually. A free version (QDA Miner Lite) also exists.

4.3.3 Some Final Notes about Qualitative Analysis

As we have mentioned numerous times throughout this book, qualitative data is often alleged to be less "real" than quantitative data. This might be due to the ease with which it can be manipulated, or how the data sample is often so small relative to large quantitative studies that it is dismissed as anecdotal. This is simply not true, and qualitative researchers can follow a number of practices to refute this perception, such as using a rigorous qualitative analysis framework, using computational tools to assist qualitative analysis, and using mixed methods of data analysis.

Also, as qualitative data is more subjective in nature, it is often useful to work with multiple analysts when analyzing it (depending on the project scope, of course). If the budget allows it, two (or more) different analysts should analyze the same data following the same methodology, which would be followed by a convergence step and discussion on divergent results and interpretation.

4.4. Obstacles to Effective Analysis

There are a number of issues a researcher has to keep in mind when doing analysis. Doing so will play a key role in conducting the data analysis in a coherent manner.

4.4.1 Confusing Facts and Opinions

When collecting data, the researcher will be faced with a number of different data sets originating from multiple sources. It is important to remember that not all of them will be facts, and should be treated accordingly A fact should be irrefutable. This can become confusing as you do online research and gather multiple statements that are in conflict: multiple different market research firms can estimate market growth differently. It will be up to the researcher to select which one is the most credible, and define it as such: use words such as "estimate" and "believe" that convey confidence, while not falsely conveying confirmation.

A fact is something that has occurred, something that is true. Do not confuse opinions and beliefs with facts.

4.4.2 Researcher Bias

Researchers are human, and can engage in research with a set of preconceived ideas and biases. It is important to be able to recognize this to be able to manage them and minimize their impact.

One of the biggest forms of researcher bias is confirmation bias, where the researcher will have an inclination to retain data that favors some preconceived ideas, and dismiss ideas that go against these preconceived ideas. Other biases can include correspondence bias (over-emphasizing a personality-based explanation for behaviors observed in others) and hindsight bias (the inclination toward seeing past events as predictable).

4.4.3 Complexity of Data

As you engage in market research, complexity can become an issue, especially if your data sources are increasing, and mixed in nature. You might find yourself in a situation where you need to reduce complexity. One way to do so is to try to simplify complex information, and distill it into a single sentence. While you might lose some of the richness of the data, you can always go back to your original data (especially if your coding framework allows quick referencing).

4.5 Future of Data Analysis

As we move forward, there is a definitive trend toward the computerization of data analysis. The emergence of software such as IBM Watson (a powerful analytics tool in its own right), will shift the way people think about and do data analysis. Speed, accuracy, and sexy visualization tools, all within reach with a couple of clicks, will change the relationship between the researcher and his data. Furthermore, as machine learning increases, we will see an improvement in basic data analytics steps such as data clean up and even predictive analysis.

Big data will also come into play. In the past, we might have looked at a data set as a unique data point. As we go forward, the emergence of big data means that the first step of analysis will be to link new data to many other sources simultaneously, trying to make sense of it in a global sense, not only the context in which it was gathered.

Furthermore, the way data is generated is shifting dramatically. Our mobile phones and devices (from a Fitbit to our credit and loyalty cards) are all individually generating data, so the future of data analysis is one where an increasing amount of data is being generated from multiple locations, and converged into massive databases.

Nonetheless, these tools should not replace the researchers' ability to interpret and read his data. The use of these tools will not teach the user basic statistics, and without these basic skills, it becomes very easy to believe the data "because my computer says so." The ability to interpret data is crucial going forward, even if tools evolve in directions that make this easier.

Chapter 5

Estimates and Models

Usually, I try to create a story with my data, highlighting an actionable insight or distilling it into a model that will help the recipient to understand my message. Hence, after collecting and analyzing the data, I like to integrate the data into a framework that eases comprehension and emphasizes the most important elements. In the following sections, we will be going over the SWOT model (and its alternative, the SCORE model), the TAM-SAM-SOM model, the Kano model, and the strategic triangle (or the 3C's).

5.1 Appraising the Market Environment: The SWOT Model

The SWOT (strengths—weaknesses—opportunities—threats) model is a simple yet clear framework used to present your organization (your product or your service) in relation to its environment. It is also possible to build SWOTs of competitors and their products.

In a SWOT model, the strengths and weaknesses refer to the organization itself (internal elements), while opportunities and threats focus on the environment outside the organization (external elements). Hence, there are three objectives to building a SWOT model: defining the organization comparatively to its competitors, identifying the best opportunities for growth, and identifying the threats (current and potential) inhibiting potential growth.

The main advantage of the SWOT model is that it is easily recognizable: most individuals have been confronted at one time or another with a SWOT model and your recipients will be able to recognize it with a quick glance, letting them capture the information you are conveying. Its simple nature lets the user condense and simplify complex data. This is also its disadvantage, as a tendency to oversimplify complex situations can leave information off the table. Another issue is that every factor is weighted equally in a SWOT model, so it is impossible at first glance to determine which factors are crucial, and which are ancillary. Finally, SWOTs are subjective, so expect the building of your SWOT model to be iterative, making changes as more information becomes available and more feedback is obtained.

Before engaging in a SWOT analysis, it is very important to clearly define the objective of the SWOT model. As we will see in the following sections, many components are subjective, but having a clear opening statement and a vision of what you are trying to achieve will help you correctly categorize your information.

5.1.1 Four Elements of a SWOT Model

The four elements of a SWOT model are divided into two broad categories: internal elements (strengths and weaknesses) and external elements (opportunities and threats). In general, internal elements are those that you have some measure of control over, while external elements are those which you have no control over.

Internal attributes are those that are unique to your organization. They are the elements that define you and differentiate you from your competitors. The bulk of this analysis is done by examining your internal resources, such as employees, board of directors, and partners, as well as financial and sales data. It could also include a scan of your organization online, identifying how your product or service is rated or perceived, as well as comparative benchmarking of leading products.

External attributes include all third parties that are outside your organization that impact your organization. Regulations, laws, suppliers, and competitors are all part of the external environment that directly impacts your organization (Figure 5.1).

Strengths are positive elements that give your organization a competitive advantage or unique value over competitors. A specific technology, an exclusive license, and a strong patent portfolio are all unique attributes that can serve to distinguish you from existing competitors. Team members with unique

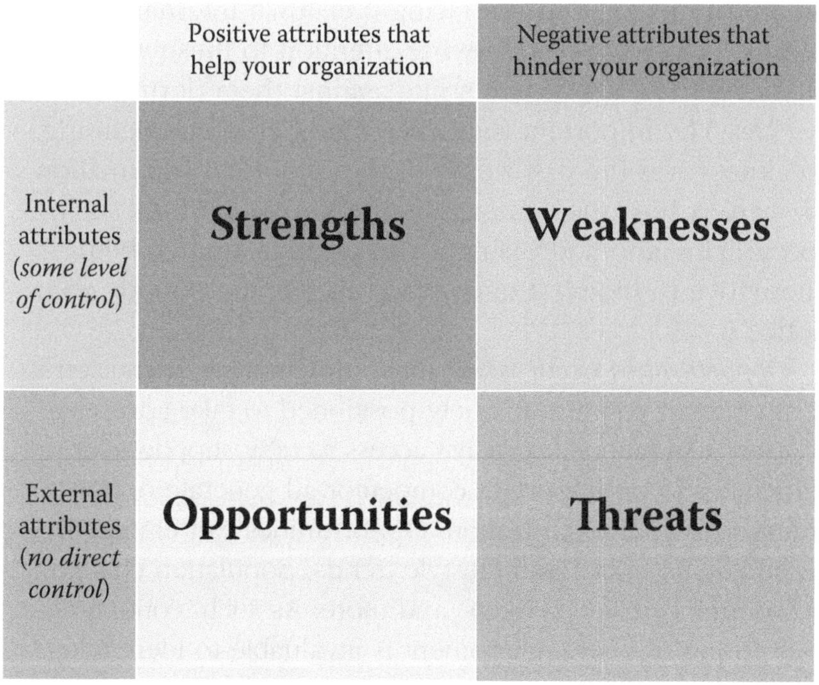

	Positive attributes that help your organization	Negative attributes that hinder your organization
Internal attributes (*some level of control*)	**Strengths**	**Weaknesses**
External attributes (*no direct control*)	**Opportunities**	**Threats**

Figure 5.1   SWOT model framework.

skills and experiences are also strengths that need to be showcased. Some of the questions that can guide you in building the Strength element of your SWOT model include: What can we do well? What is our unique expertise? What unique resources do we have access to? What advantages do we have over our competitors?

Weaknesses are the attributes that are inhibiting growth in your organization. They are the attributes that place you at a disadvantage compared to your competitors. For example, using a legacy production system to build your product, lack of sales expertise, or not having a patent to protect your innovation are all weaknesses that can inhibit your growth. Some questions you can ask yourself when defining your weaknesses include: What is it that we are not good at doing? What are our competitors doing better than us? Is there anything we do that we know we could improve?

Weaknesses can be difficult to define and admit. Many organizations are wary of facing their own internal weaknesses, or believe that drawing attention to these weaknesses will be a potential strike against them during investor pitches. The important thing to remember is that defining weaknesses is the first step, but the important information to share is how the company will address them. As such, recognizing and addressing a weakness is a much better alternative to hiding it under the rug, hoping nobody will notice it.

Opportunities occur when there are changes in your environment that you are uniquely positioned to take advantage of. New regulations, exclusive access to new suppliers, or the difficulties of an important competitor all generate opportunities for your organization. Opportunities can emerge in technologies, social and lifestyle trends, population changes, government policy, markets, and more. As such, continuous monitoring of your environment is invaluable to identifying opportunities. Some questions that can guide you in identifying opportunities include: What are some of the trends impacting my client base? What are some of the major new regulations that have impacted my industry in the last few years?

Threats are shifts outside your organization that can have an impact on your organization. While they are outside your organization (meaning you have little control over them), recognizing and assessing them can help you prepare your organization, and identify solutions to address them. For example, a shift in regulatory procedures could incentivize you to change some elements of your production model, and look for appropriate subsidies and partners to help you make the necessary adjustments.

You might not be able to change the external environment, but you can react once the change occurs (or even better, you can plan ahead to make adjustments before the change occurs). Some examples of threats include new competitors entering your market, current clients shifting their strategy, new technologies affecting consumers' purchase and use of your products, changes in social trends, government regulations, or changes in the availability of crucial supplies. Some of the questions you can ask yourself are: What are my competitors doing? What new technologies are being developed? What are some of the regulatory components that are under review or discussion?

5.1.2 Developing Strategy Applications from Your SWOT Model

There are two simple approaches to developing a strategy using the information in your SWOT model: matching and converting.

Matching implies the pairing of your positive attributes (strengths and opportunities) to identify prospects for your organization. For example, if one of your strengths is a unique production process, and an opportunity emerges in lower regulatory hurdles for a specific market, you might then identify that market as a potential new market to target with your product as you are able to scale up rapidly.

Converting means identifying a weakness or threat and transforming it into a strength or opportunity. For example, if a competing start-up is developing an innovative technology that threatens your market, this might be an interesting opportunity to partner up and co-develop said technology. Other opportunities could be developing a similar innovation or partnering with a competing innovative technology as a way to develop a new *unique selling proposition*.

5.1.3 An Alternative to SWOT: The SCORE Model

An alternative to the SWOT model is the SCORE model. Like SWOT, it is a strategy framework used to develop a quick assessment of your current situation. SCORE stands for **S**trengths, **C**hallenges, **O**ptions, **R**esponses, and **E**ffectiveness. SCORE is less of an environmental scan tool, but rather a problem–solution model. You identify your strengths and the challenges you are facing, and start developing solutions to address them. As such, it is more of a road map toward problem solving than SWOT is, and includes a measurable component to ensure you are on track.

The key to using SCORE is to select an issue and start developing the framework. Assess each item as you progress, and identify any measurable items.

Finally, remember that while SWOT is somewhat static, SCORE is an iterative model, and can be re-examined on a recurring basis. It is a useful model to demonstrate the acknowledgment of issues, as well as a road map to resolution. To assist you in building your own SCORE model, you can use the SCORE framework development table included (Table 5.1).

Table 5.1 SCORE framework development

SCORE Attribute	Question Guide and Objective
Strengths	• What are my current capabilities? • What are my current internal resources? • What are some of the outsides resources that I can call on? *Objective: Inventory of capabilities*
Challenges	• What are some of the issues I need to address? • What are the capabilities/resources I need to acquire/develop? *Objective: Inventory of issues*
Options	• What are the options to address these challenges? • What is the risk associated with each option? • How should we prioritize them? *Objective: Strategy scenarios*
Responses	• What are the expected responses to our options? • From regulators? competitors? stakeholders? • What return/reward do we expect from each option? *Objective: Risk-management scenarios*

Effectiveness	• For each option
	  • Is it effective?
	  • Is it realistic?
	  • Is it reliable?
	Objective: Project management assessment

5.2 Appraising Market Size: The TAM-SAM-SOM Model

To appraise your market size, a TAM-SAM-SOM model is an interesting approach since it combines elements relating to both the top-down market sizing approach (sizing the market size from the literature) and bottom-up approach (estimating the market size from the perspective of the organization's available resources).

There are multiple advantages to this model. First, it is relatively easy to use and explain. Second, it combines the top-down and bottom-up parts of market research, which are essential to telling your story. Finally, a well-built TAM-SAM-SOM demonstrates a solid understanding of the market and its limits.

On bottom-up market research ...

One of the main issues I see when going through investment pitches with start-ups is the lack of bottoms-up data to justify their market size. If your plan is to reach 5 million dollars in sales with two sales people, at 1,000$ a unit, does that mean that they are selling annually 2,500 units each, or over 50 units a week? Is that really possible? Without a good bottoms-up approach to market sizing, it's often very difficult to give credibility to an otherwise very good presentation.

Ajit Singh, Managing Director and General Partner,
Artiman Ventures

The main issue with TAM-SAM-SOM model is that it relies on the strength of your assumptions. As such, building a solid model will depend on a realistic assessment of the market and of the capabilities of your organization. Furthermore, the justification behind your commercialization plan will be crucial in justifying your SOM.

5.2.1 Three Parts of the TAM-SAM-SOM Model

The TAM-SAM-SOM model is divided into three distinct parts that are imbricated into one another. The largest part is

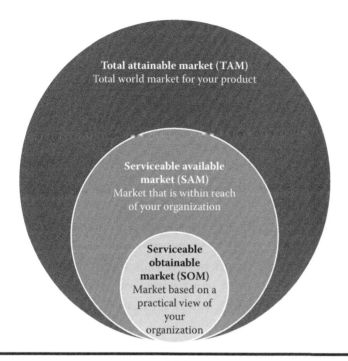

Figure 5.2   TAM-SAM-SOM model framework.

the TAM, which is the *total attainable market* for a specific product or service. It contains the SAM, which is the *serviceable available market*, the market that is within reach of your company. This in turn contains the SOM, which is the *serviceable obtainable market*, the realistic market share that you are likely to obtain, or the one that you plan to achieve in the short term (Figure 5.2).

To properly illustrate, let's use the very simple example of a small biotechnology company developing a diagnostic device in the small cell lung cancer segment.

■ The *total attainable market* is the market from its broadest perspective. For our small start-up, the TAM is the total lung cancer diagnostic market space. In this regard, Grand View Research estimated that the global lung cancer diagnostics market revenue is expected to reach

$3.6 billion by 2024.* This could be seen as an approximation of the future total available market worldwide. But a company can seldom claim this to be their target market. The resources necessary to target this total market simultaneously are quite ambitious. So, the next step is to calculate the serviceable available market.

■ The *serviceable available market* is the area that is within reach of your company. For example, in the case of our small start-up, we can make a series of assumptions:

a. The diagnostic device specifically detects small cell lung cancer, which accounts for around 15% of the lung cancer diagnostic market share.

b. In the short term, your company will be targeting mainly the North American market, which accounts for 30% of the revenue share.

Hence, your serviceable available market share is estimated at $162 million (Table 5.2).

Table 5.2   Sample Serviceable Available Market Development

Estimated TAM	3.6 billion
Estimated market share of small cell lung cancer segment	15%
Estimated market share of North American market	30%
Estimated SAM	162 million

---

* Grand View Research. Lung Cancer Diagnostics Market Worth $3.64 Billion by 2024. June 2016. https://www.grandviewresearch.com/press-release/global-lung-cancer-diagnostics-market (Accessed January 28, 2017).

■ Finally, you start looking at your *serviceable obtainable market*, which is the realistic market share that you can likely acquire, or the one that you plan on achieving in the short term. This data has to be backed by a bottom-up forecast, detailing how you will utilize resources to achieve your revenue goals. Without going into the details of a commercialization strategy, let's imagine that you have made a two-year plan that generates $10 million in sales.

Once you have developed your model and data, it becomes quite easy to demonstrate your market with a bubble graph as shown in Figure 5.3.

As you can see from Figure 5.3, this type of modeling depends on the series of assumptions. Your market research will be the cornerstone of your results. A good understanding of the market and the competitors is essential to building

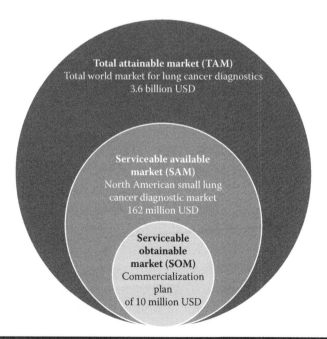

Figure 5.3   TAM-SAM-SOM model for a small cell lung cancer diagnostic device.

a justifiable SOM, and recipients will be as interested in the numbers as how you developed the model: was the presenter realistic when he built his model, was he pragmatic? Remember that, in general, forecasting a SOM that is greater than 10% of

---

On the importance of realistic assessments ...

Many companies overestimate the ability of their therapeutic to penetrate a market with an approved drug on the market. Oftentimes, we see companies overstate patients' willingness to try an effective, but highly invasive therapy over a less effective, but more tolerable standard of care product.

Caroline Stout, Investor at EcoR1 Capital

---

the SAM is going to require a great deal of justification.

Also, as with everything else, TAM-SAM-SOM models are iterative, and can be adjusted as new data becomes available. Finally, it's always a good bet to be a bit more conservative with your numbers: it's easier to explain how you beat your forecasts than justifying why you fell short of the number that you promised.

5.3 Appraising Customer Attributes: The Kano Model

Your customers' appreciation of your product attributes can be classified following many different models. In this section, we will be looking at the Kano model of attribute classification. The Kano model classifies purchasing attributes into five distinct categories: threshold attributes, performance attributes, excitement attributes, neutral attributes, and reverse attributes.

The Kano model is a useful framework to classify customer perceptions, and enables readers to quickly identify which features will generate the higher returns (since they are linked to creating higher satisfaction), and which features are not worth investing in (since they are not creating any satisfaction, or even worse, they are generating dissatisfaction). Take note that these attributes do migrate over time. The longer an excitement attribute has been promoted, the more it becomes a performance or threshold attribute. We will be going over each attribute, followed by a short discussion on the limits of the Kano model, which includes the dimension of time.

5.3.1 Five Attributes of the Kano Model

As mentioned in Section 5.3 there are five distinct purchasing attribute categories in the Kano model: threshold attributes, performance attributes, excitement attributes, neutral attributes, and reverse attributes.

Threshold attributes are the attributes that the consumer expects in a product. They must be present, and since the consumer anticipates them, they don't usually create a differentiation opportunity. Increasing the focus on or investment in these attributes does not usually provide more sales or greater client attraction, but removing or diminishing these attributes will result in extreme customer dissatisfaction. In other words, the customer expects that when purchasing the product, their basic needs will be fulfilled, and that this attribute will be present to fulfill that need. A simple example of a threshold attribute is the absorbent pad on an adhesive bandage. The client expects that there will be an absorbent pad on the bandage; the lack of a good absorbent pad on an adhesive bandage will turn clients away, and it would be difficult to focus on this pad as a unique selling point.

Performance attributes are those where the more the attribute is present, the better it is for the consumer and it will improve customer satisfaction. They are clearly expected by clients. The price a customer is willing to pay for a product is closely tied to performance attributes. Nonetheless, products that focus on performance attributes are interchangeable with other products focusing on performance attributes, so it is unadvisable to build a product strategy that focuses exclusively on performance attributes. A performance attribute in an adhesive bandage could be a material that makes the consumer more comfortable or a material/glue that makes the adhesive stay on the finger longer.

Excitement attributes are those that are unanticipated by customers, but which create high levels of customer satisfaction. However, their absence does not generate dissatisfaction. These are the attributes that enable you to differentiate your product from products with similar performance attributes. Providing excitement attributes that address "unknown needs" generates a competitive advantage over competitors. Following our adhesive bandage example, bandages with antiseptics directly in the absorbent pad or bandages that help to regenerate wounds could be examples of excitement attributes for consumers.

Neutral attributes are attributes for which the customer does not manifest a distinct opinion. Their presence neither improves nor impedes the purchasing decision. Frequently, these are features that excite the innovator, but are seldom of interest to customers. Hence, it doesn't mean that the attribute doesn't add value to the product, it just doesn't add any value for the consumer. For example, the color of the adhesive generally does not influence a purchasing decision (unless you have chosen an especially hideous color, in which case you might have unknowingly generated a reverse attribute), but it might be easier for a manufacturer to privilege and produce their product in a single color.

Reverse attributes are rather rare, and occur when a client perceives an attribute negatively. In that case, he will be willing to pay an amount of money to have them removed. As such, the presence of a reverse attribute creates a negative experience, while its absence creates a positive experience. For example, some clients might be willing to pay more for plain adhesive bandages, rather than buying cheaper adhesive bandages with kids' cartoons on them.

The Kano model is a simple but powerful model, useful in identifying opportunities in customer needs, as well as demonstrating how these attributes impact the consumer's purchasing decision. The more the attributes create customer satisfaction, the more effort should be invested in developing and refining these attributes.

5.3.2 Issues with Kano Modeling

There are two issues with Kano modeling. One is the impact of time on attributes, the other issue relates to different consumer perceptions.

The main issue with Kano modeling is the impact of time on the perception of consumers. For example, an attribute that was initially classified as an excitement attribute can quickly become a threshold attribute as more and more companies integrate the target feature. For example, think back to a few years ago when cameras and MP3 players were integrated into mobile phone devices. While this was once an excitement attribute (*Wow! I don't have to carry three devices anymore, everything fits onto a single device!*), could you now imagine purchasing a mobile device that couldn't play music, or couldn't take pictures? As such, the Kano model needs constant reappraising to ensure that attributes are correctly classified, and that the market has not shifted away from some of your previous assumptions.

The second issue is that you can find inconsistencies among consumer perceptions. As you start collecting data, you might realize that while some attributes are a threshold attribute for some individuals (a camera is a must for some potential clients), it becomes a negative attribute for others (*I have to pay extra for a camera? But I never use it! Show me a cheaper model without a camera instead!*). Careful segmentation of your data and client targeting alleviates this issue: as you precisely identify your customer segment, you will eventually distinguish patterns between them. The clearer your end market, the less likely you will have inconsistencies in consumer perceptions.

5.4 Appraising Competitive Space: The Strategic Triangle (3C's)

The 3C's is a simple framework that focuses on the immediate competitive environment, which includes your Company, your Competitors, and your Customers (Figure 5.4). It is used to establish your competitive position relative to your customers and competitors, and is useful for short-term planning.

The premise of the 3C's is that your competitive advantage is generated by your ability to create greater value than your competitors. As the inventor of the 3C's wrote, "A successful strategy is one that ensures a better or stronger matching of a corporation's strengths to customer needs than is provided by competitors" (Kenichi Ohmae, 1982).

Developing the 3C's comes with certain assumptions. First, you must be in an environment where suppliers do not possess enough power over you to influence your company's strategy. Second, substitutes and new entrants are included in competitors, rather than being positioned as a separate entity. Finally, there must be some level of environmental stability to let you focus on the three entities.

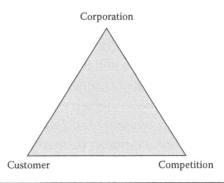

Figure 5.4   The 3C's model.

5.4.1 Implementing 3C's Strategic Triangle

There are four steps to using the 3C's strategic triangle to develop an integrated market strategy. Sequentially, decisions are made for each attribute, and then an integrated solution is developed that encompasses all three decisions.

First, define your *customer-based strategy*. It is recommended to put your customers at the heart of your 3C's strategy. Putting your corporation first is especially difficult, as it will create a bias throughout the rest of the analysis as you keep benchmarking elements against yourself. Your company should identify a segment of customers and focus on them, instead of trying to satisfy the entire market. Based on your market research, you will be able to develop your key market segments, the customers whose needs you can cater to. Identifying their concerns should be the basis for your strategy.

Second, define your *competitor-based strategy*: You have to identify areas where the competitive situation allows you to develop a clear advantage. While you can have many competitors, try to restrict yourself to your top three to five most direct competitors. This will eliminate the risk of muddying the water too much prior to the analysis. Based on your competitive review, you might already have identified space where you have a better advantage such as a superior technology, or a stronger capital base, for example.

Thirdly, define your *company-based strategy*. Identify your key functional attributes and core competencies. Choose which competencies you want to focus on, and make the decision on which functions you want to outsource to specialized third parties.

Finally, work to unify all of these individual competitive strategies to create a single coherent strategy. Due to their natural interdependence, this should enable you to create a single strong strategy.

5.5 A Final Note on Organizing Complex Data

This section is dedicated to describing a few formal and well-known models to help organize your data. They don't apply to every situation, and there are many others we didn't go over that could be of use, such as Porter's Five Forces Model for competitive strategy or the PEST (political—economic—social—technological) analysis for understanding how outside environmental factors affect the organization.

Nonetheless, when organizing data for presentation, there are a few things I try to keep in mind. First, when reviewing my data, I look to make sure that it is logically structured. Look for opportunities to showcase symmetries (these are usually well understood) and make sure to showcase clear connections and flow. Also, take a look at your data, and try to identify any unimportant data that you can remove.

Afterwards, take a step back to look at the big picture and the details to see if the point you were trying to make came across clearly. It can be useful to ask someone else to review your data, just to make sure that the point is consistent and clear, so ask for a second opinion. The collaboration that results enriches the presentation, and produces better overall analysis.

Finally, remember to focus on your message. It is quite easy to lose your main message when analyzing complex data, so make sure to compare your final results to your initial message to ensure there is a match between them.

Chapter 6

The Look n' Feel

Think back to presentations you have attended in the past. I am sure you can recall some pretty boring ones, with traditional PowerPoint slides that obfuscate information, rather than communicate it. It might have been that the slides were made of a "solid wall of text" (a dozen or more full sentences written in a 16 font), which the presenter promptly read like he was reading a chapter from a book. It might have been the color contrasts that made the text illegible, spelling mistakes that made the text unreadable, or badly labeled graphs that were impossible to decipher.

These presentations failed, not because of the information, but rather because the people presenting them failed the *look n' feel* of their presentation.

When presenting, the look n' feel of your data is a crucial element to ensure that the audience understands the message. Mishandling the presentation elements can drown you lose your message, and the intended recipient will not receive the information you are attempting to share the way that you meant them to. Even worse, the recipient could interpret it completely differently to the way you were trying to convey it. Presenting data with attractive and clear visuals is essential to sharing the results of your data collection and analysis, as well as helping you make better-informed decisions.

Transforming data into eye-catching graphics and tables is not as intuitive as it may seem. While new software and platforms make the mechanical aspects of data presentation simple to master, everything from choosing the right graphics to the graphical design aspects are a unique set of skills, skills I refer to as look n' feel skills. They include choices related to the look of the presentation (colors, fonts, shapes, layout, graphics) and the feel of the presentation (architecture of the graphics and tables, responsiveness, and interactivity). Look n' feel also includes the effort to create a consistent image across a presentation, and the overall branding image that is generated.

This chapter will be dedicated exclusively to communicating our market research findings to our audience. The key to a good graphical presentation is to select the method that best fits the data, so we will start by going through a process to choose the best way to communicate quantitative information. This will be followed by information on models for presenting qualitative data. The chapter will conclude with an overview of presentation tools that integrate quantitative and qualitative data, such as slideshow tools, the Prezi platform, and the use of infographics.

6.1 Presenting Quantitative Data

The transformation of quantitative data into comprehensible output is a necessary step to getting your message across. To do so, it is important to decide which graphical layout is the best one to demonstrate your information. We will be going over a simple two-step process to enable you to quickly identify which is the best graphical layout for your data.

6.1.1 Identification and Evaluation Step

Following data analysis, the first step to communicating your data is to identify your message, followed by a short evaluation stage between a table and a graphical layout.

Identifying your message means determining what it is you wish to communicate with your data, and which elements you will choose to emphasize your message. There are probably many angles you can develop with your data, so it is important to carefully craft the message you want to share, while removing information that isn't important as unnecessary information might obscure your message rather than enhance it. Looking back to your research objectives will help you craft that message, but overall, the data you have collected will more than likely already produce the best stories for you to share.

Once you have crafted the message you want to share with the data, *evaluate* whether a table or a graphic is needed to share the message.

■ *Tables are used when direct access to precise numbers is necessary to support your message.*
In a *table*, information is displayed across rows and columns. A table is useful if precise individual values have to be shared. For example, imagine doing some market research to identify the geographic market share of a drug: if sharing the precise market share percentage for each geographic region is crucial to the presentation, the use of a table is indicated. Ideally, a table should have a clear title, include information such as the sample size and when the data was collected, and be inserted after the text that it relates to.

■ *Graphs are indicated when a relationship in data is what you are trying to demonstrate.*
A *graph* is data that is displayed in a visual layout, such as a pie chart, a line graph, or a histogram. Visualized

data allows you to demonstrate information in relation to another data set, along one or more axes, and is useful when you are trying to show the shape of data, its patterns, and its trends. Following the previous example, if the message you want to circulate is how much of the drug is sold in a geographic area in relation to other competitors, then the use of a chart (such as a pie chart) effectively communicates relative market size from one company to another. Ideally, a good graph will include a title, a clear scale, and clear axes so the reader can quickly integrate the message.

Note that if you are showcasing a single important point of information or one data point, you should simply use large tiles or labels for your data, rather than embedding it in a table, a graphic, or a visual component.

6.1.2 Transforming Quantitative Data into a Graphic

If you have determined that a visual layout is needed for your information, you can identify which graphical representation is the most pertinent to communicate your message. Remember that "fancier" does not mean better. Sometimes, a simple line chart is all you need to properly convey the information you are trying to share.

A simple approach to selecting which is the best visual layout for your data is to identify the data relationship you want to demonstrate. Once you have decided the type of relationship you want to profile, you can use Table 6.1 to identify the best graphic representation for your data.

As a broad example, a few years ago, I wrote an article for a marketing magazine.* Using data from a client's

---

* Denault, J. 2014. How much do environmental attributes influence purchasing patterns? Marketing Magazine. https://www.marketingmag.com.au/hubs-c/how-much-do-environmental-attributes-influence-purchasing-patterns/#.U4MYzfldXL8 (Accessed January 10, 2017).

Table 6.1 Choosing the Most Relevant Graphic Representation for Your Data According to the Type of Relationship

Relationship	Description	Graphic Representation Alternatives
Time	You are observing a single variable over time	• Use a *line chart* to emphasis the shape of data • Use a *bar chart* to emphasize comparison between data sets
Ranking/score	You are grading items in order during a single period	• Use a *bar chart*, especially horizontal bars, when you have a long label to place • Use a *radar chart* if you are comparing scores across multiple objects
Part of a whole	The items you have analyzed are part of a whole	• Use a *pie chart* when comparing one key item versus the whole market • Use a *bar chart* when comparing trends across attributes is important • Use *stacked bars* when comparing whole categories across time
Benchmarking	Items are compared to a reference item	• Use a *bar chart* to compare your data to benchmarks • Use a *line chart* to emphasize the shape of data
Frequency	You have compiled a number of observations per interval	• Use a *histogram* to emphasize individual values
Correlation	You are comparing two (or more) variables across time	• Use a *scatter plot* to show how one attribute is affected by a second variable • Use a *bubble* to show data from a scatter plot, emphasizing a third variable

quinquennial survey, I wanted to explore the relationship between a product's environmental attributes and consumer purchasing patterns, as well as measuring how these patterns evolved over time. The hypothesis was that over time, environmental attributes would become more important in a consumer's decision process.

However, the data articulated a completely different story, as there was no significant difference in consumer purchasing patterns between 2009 and 2013. The importance of environmentally friendly attributes and the willingness to pay more remained at the same level between the two surveys. As I was demonstrating multiple relationships in the data (multiple time periods vs. purchasing interest), I used a simple bar graph to illustrate the data. This way, the reader could clearly see that the consumer levels remained constant from one data set to the next. The use of a graph was indicated because (1) I was showcasing a trend, and (2) the relationships between time and attributes were the key to understanding the story (Figure 6.1).

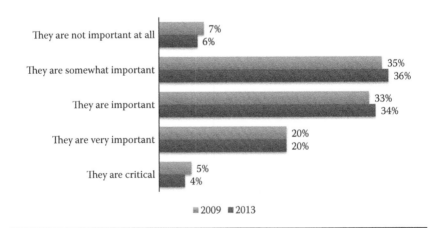

Figure 6.1   Using a bar graph to illustrate "How much influence environmentally friendly attributes have on your purchasing decision."

6.1.3 Building a Graph

Once you have identified which type of visual layout you wish to build, and have started the design, you can use the following road map to be sure that you are not forgetting any information.

1. *Before building the graph*
   ■ If you are doing a graph that mixes quantitative and qualitative scales, place the quantitative scale on the y-axis (vertical) and the categorical scale on the x-axis (horizontal); time is especially important to place horizontally since it follows our intuitive habit of seeing time moving from left to right.
   ■ If you use a bar chart, your horizontal scale will most usually start at zero, but if you are using a line chart, you can narrow the scale and start the data at a more relevant number. This will give the reader a better sense of the pattern or story you are trying to convey.
2. *Once you have built your graph*
   ■ Remove any distractions that have crept into your story. When building your graph, focus on the data itself, and try to remove any graphical elements that aren't helping the reader to understand the message you are conveying. For example, the use of a grid line is not recommended unless it is useful for helping the reader to interpret the data (e.g., the reader needs to see precisely where the data is.)
   ■ A note on color coding: Use softer natural colors when building graphical elements, while saving more flashy colors for items you need to stand out. Once you determine a color scheme, stick to it for the entire presentation, and do not shift from one color scheme to another as the presentation advances, as it becomes an additional distraction. More information on colors is available in Section 6.3.1 on slideshows.

■ Do not forget to include all relevant descriptive text: The graph title and the axis titles are invaluable for the reader to understand the message you are communicating, as well as the legend that further contextualizes the data.

■ Decide if you need to emphasize a specific data set. If one element of data is more important than the others, highlight it accordingly, using either bold colors or a different graphical item.

6.1.4 Decision Tree Modeling

Decision tree modeling is a visual presentation model used to illustrate a decision process and its corresponding actions. It can be used to illustrate any process with choices such as a purchasing process or the impact of a marketing campaign. The top of the tree refers to the choice alternatives, followed by the decision criteria, and closing with the decision outcome.

As an illustrative model, decision trees have many advantages. First, they are easy to understand, as they are a visual representation of a phenomenon, and they are very intuitive to follow. They are also very useful to enhance decision making, or as a starting point to brainstorm. For example, a decision tree could be used to illustrate a consumer decision process, and to help make decisions going forward. It can also illustrate how one data set impacts another. Finally, it requires less work around data cleanup, and can be done fairly early in the data analysis phase.

The main use of a decision tree in market research is to schematize the customers' purchasing decision process. Once in hand, it can help the company understand which factors are the most important in their purchasing decision, and once properly illustrated, can be used to emphasize a product's key differentiating factors.

Coming back to an earlier example presented in Chapter 4 (Section 4.2.2.1), my client in the dental market had developed a new combination of product and service, and was interested in determining how much the delivery model impacted consumers' interest to purchase, relative to the price he was targeting. As such, a decision tree was modeled to help illustrate the collected data (Figure 6.2).

In the decision tree, the choice was the purchasing decision at a certain price point, the decision criteria was the delivery speed, and the outcome was the purchasing decision. As we can see from Figure 6.2, the slower the proposed delivery model, the lower the purchasing intent.

Overall, decision trees can provide insight into the decision process, enable better decision making, and illustrate to the audience the data behind a specific decision.

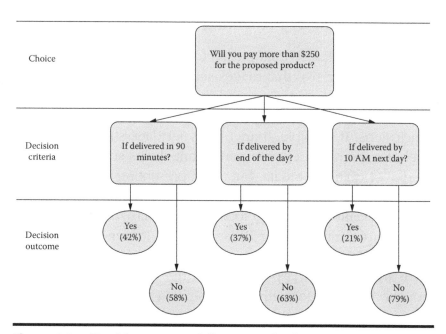

Figure 6.2   Decision tree modeling example showcasing how price, delivery model, and purchase intent are interrelated.

6.2 Presenting Qualitative Data

Qualitative data differs from quantitative data during analysis, a difference that persists all the way through to the presentation of data. The richness of qualitative data is descriptive in nature, which does not necessarily lend itself to graphical representations. A literature review of close to 800 scientific articles found that only 27% of articles that presented qualitative data used classical graphics to present the information.* Hence, presenting qualitative data focuses on the narrative and the story. To do this, we will be going over a few tools that can be used to either reinforce the presentation, or to help illustrate the collected data.

Of note, if you have coded your data into a quantitative format, feel free to use the different tools and visual representations that we discussed in Section 6.1. Codifying qualitative data is reviewed in depth in Section 4.3.1.

6.2.1 Overview of Presenting Qualitative Data

Presenting qualitative data means focusing on organizing the story into a narrative that makes sense to you and the readers. Hence, focus on the general themes that you have identified during data analysis, and focus on how you got to these conclusions. If possible, triangulate your data: showing the level of consensus between participants helps support the validity of qualitative data, but avoid presenting elements in a way that overemphasizes the quantitative dimensions of your data, especially if you used tools that emphasized exploration. For example, interviewing 10 key opinion leaders will generate valuable information, but you should never use a quantitative argument (i.e., 4 people out of the 10 we interviewed thought that ...) to formally demonstrate a point. This type of

---

* Verdinelli S, Scagnoli N. 2013. Data display in qualitative research. *International Journal of Qualitative Methods*. Volume 12: 359–381.

demonstration tends to bring out counter arguments focusing on sample size, rather than embarking on the narrative you have built.

6.2.2 Using Quotes to Reinforce Qualitative Presentations

A presentation based on qualitative data should reflect that the data consists of patterns found in quotes and observations. Hence, it is quite common to use direct quotes to reinforce conclusions during presentations. Quotes can also be used in reports to reinforce quantitative data, as a way to strengthen an element that is being emphasized, or that was discovered during analysis. If you choose to directly quote someone in a presentation or a report, remember to respect confidentiality and attribute quotes anonymously (unless express permission to directly quote is given).

Also, while you can use some brief quotes to illustrate a point, don't use too many in the same section of your document. Avoid over-quoting participants as a way of delivering qualitative data: quotes illustrate the results of the analysis, but do not replace it. Stringing a series of quotations together without analysis makes it difficult for the reader to follow (especially if the narrative style shifts from one quotation to the next).

Finally, it is tempting to choose the most remarkable quotes from your interviews as a way to demonstrate a point, but remember that these do not necessarily represent the pattern of your data. Choose quotes that represent a concerted point of view, and avoid outliers as a way to illustrate a point.

One way to emphasize a quotation is to use a *boxed display*. Boxed displays are text framed within a box. These are used to highlight a specific narrative or quote that is important enough to be put into evidence. It emphasizes something of specific interest, and helps separate a concept and a supporting quotation. As you might have noticed throughout this book, I have used box displays to illustrate some important points that were mentioned by the various people I interviewed. When I did so, I used a consistent presentation style. First, I put in a centered bolded title. The text is normal, no need to be justified. Following this, the signature is in italics, right-justified. Finally, I use a shaded square, with a line contour, both side indented a bit as a way to draw intention to the quotation. The results looks like the following quotation (Figure 6.3):

> **On the importance of boxed displays ...**
>
> A few years ago, I completed a project with a client in South-East Asia. The project was to identify the feasibility of establishing a small bio industrial complex, based on regional strengths and competition. The report included a number of interviews with companies interested in the region, but these interviews had revealed a number of potential issues. To ensure the issues were noticed by the reader and that they would be addressed, boxed displays (such as this one) were used sparingly throughout the report, emphasizing those latent issues.
>
> *Jean-Francois Denault, Impacts. Ca*

Figure 6.3 Use of a boxed display to share qualitative information (quotes).

6.2.3 Visual Layouts to Display Qualitative Data

The imaginative use of diagrams and schematics can be useful ways to illustrate analytical processes and findings, and to simplify more complex information. While there are many ways to visually display information in qualitative presentations, we will be emphasizing matrixes and flowcharts.

6.2.3.1 Qualitative Matrixes

Matrixes are built by crossing two or more dimensions, variables, or concepts of relevance to the topic of interest to see how they interact. They are used to classify data across topics and demographic data, or provide a more complex illustration of results. The advantage of building a qualitative matrix is that when you are building it, you get to understand your data further, giving you an additional level of analysis. Of course, not everyone is visually oriented, but they are an interesting way to display information.

To illustrate, let's refer back to an earlier example I mentioned in Chapter 4 (Section 4.3.1.2), a survey done for a media client active in the Generation Y demographic space. For example, in this case, we can notice how the importance of health as a purchasing factor is slightly higher in women than in men. This client was investigating purchasing patterns of healthy drinks. He had asked participants about the last drink they had purchased, and what had made them purchase that item over another competing brand or product. Since the objective was purely exploratory, the answers that participants gave were open-ended. Once the data was codified, I created a framework with over a dozen different categories of responses. The top five answers had the most responses. To see the difference between purchasing factors (qualitative data) and the sex of the person answering the survey (quantitative data), we built a matrix (Table 6.2).

Table 6.2   Example of a Mixed Matrix (Qualitative and Quantitative Data) Showing the Relationship between Purchasing Factors and the Sex of the Respondent

	Sample	%	Female	%	Male	%
Taste	689	68.9	336	67.2	352	70.4
Price	366	36.6	182	36.4	183	36.6
Refreshment	278	27.8	137	27.4	140	28.0
Health	271	27.1	156	31.2	114	22.8
Availability	239	23.9	117	23.4	121	24.2

The advantage of a matrix is that it has the feel of quantitative data, and is useful to convey messages, especially when you have access to large data samples.

6.2.3.2 Flowcharts

Flowcharts are diagrams used to document, analyze, and illustrate a process or organizational structure using various graphical elements, connected by arrows. They help visualize what is happening, can help illustrate a problem, and can provide a possible solution. Also, they can be used to effectively communicate a process to other parties so they understand it. In presentations, flowcharts are useful to illustrate an existing process, for which a product or service resolves an issue or solves a problem. They can be a useful way to illustrate qualitative data you have gathered from interviews (e.g., how stakeholders usually perform an activity, or how they complete a transaction), and to triangulate it with other sources of information (such as secondary research or observation).

Some conventions around flowcharts include the use of rectangles when illustrating an activity, and a diamond when illustrating a decision. Also, stay consistent throughout your flowchart: a consistent approach eliminates unnecessary distractions and lets the reader focus on the essentials. Try to keep shapes the same size as much as possible, and limit the use of multiple colors. Finally, try to keep everything on one page: if it gets too crowded, it might be an indication that you have multiple processes or activities intertwined, so try splitting them up into two or more flowcharts, linking them as needed.

When building a flowchart, start by organizing the tasks in chronological order of when they happen, or following a hierarchical structure. Note when decision points occur, and what the various consequences are. Also, try to take note of when there is a feedback loop (an activity that returns to a previous step in your flowchart). Then, do a first draft of your chart: I use a whiteboard as first drafts are often cluttered, and redundant steps may need to be removed. Once you have a stable flowchart, you can use Word or Excel to illustrate it, or a more specialized software program or website if you are building something particularly complex.

To illustrate, I refer back to a research project I did quite a few years ago. A chief executive officer (CEO) of an emerging start-up was interested in benchmarking personnel assigned to clinical research per product in biotechnology firms in Canada and the United States. Since information in this field is usually proprietary (and not available at large), research was done through three indirect sources: a review of industry literature, interviews with individuals, and a review of the hiring history for companies of a similar size following a similar business model. The hypothesis was that by triangulating three independent sources, we would obtain accurate data.

While the final results are proprietary, we are able to share an example of a flowchart that was built from our consolidated data. Already, this gave the CEO some perspective on organizational development and trends (Figure 6.4).

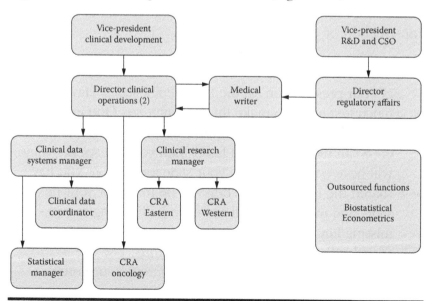

Figure 6.4   Hypothetical clinical development structure for mid-sized biotechnology company.

6.3 Presentation Tools

There are some interesting presentation tools that you can use to present information, especially if you are consolidating different visual layouts, models, and information into a coherent story. First, we will take a moment to go over some slideshow basics (including the Kawasaki 10–20–30 method), followed by an overview of visual storytelling software and infographics.

6.3.1 Slideshow

Slideshows are presentations consisting of information, pictures, videos, and audio shared in a sequential order using an electronic device. They have the advantage of being (potentially) more attractive and interesting than a traditional paper document, and can be exported in standardized file formats (such as a PDF) so they can be distributed at large.

While we assume that most people reading this book will be familiar with the basics of doing a slideshow, there are a number of fundamentals that you should remember during the construction of your deck.

1. *Use the least text possible*: Do not put every single word of your presentation on the PowerPoint slides. Having too much text tends to make the presenter *read* the information, rather than *present* the information. As your audience can most likely read faster than you can speak,* they will have completely read your slide long before you have shared the information. They will be ready for your next slide, and most likely will not even be listening to you, missing key information or drifting away from your content.
2. *Standardize the fonts and backgrounds*: Use the same fonts (sans serif, if possible) across your entire

---

* If you speak faster than attendees can read your presentation, you possibly have another presentation issue on your hands.

presentation. Use one font size for your titles, and one font size for your text. A third font size can be used only if you want to specifically emphasize some key information. Standardized fonts and backgrounds make it easier for the audience to follow without being distracted. An alternative to standardized fonts is using a font matcher, such as FontPair (http://fontpair.co/), to find proven font combinations for both your presentations and your documents.

3. *Use the 6×6 matrix*: Synthesize your information as much as possible. Try to keep every slide following the 6×6 matrix: a maximum of six bullet points per slide and six words per bullet point. The objective is to make sure that you have synthesized your information, and that you own your content beforehand. This has the added benefit that you won't be able to read your slides mechanically.

4. *Test your presentation on-site*: If possible, arrive early and test your presentation on the computer provided, as well as testing the projector; there is nothing more destabilizing than having technical issues or a formatting glitch midway through a presentation.

5. *Ask your audience to hold questions until the end of the slide*: Some specialists advocate that the audience should wait until the end of the presentation to ask questions, but there are some issues with this approach. The main problem is that by the end of the presentation, your audience will most likely have forgotten their question. Worse, some will struggle to remember their question all through your presentation, focusing on it and not listening to new information as you share it.

Other presentation specialists suggest that people in the audience ask questions as they occur, to the point of "feeling free to interrupt" the presenter when they have questions. This is also not something I recommend, as it constantly interrupts the flow of your presentation, and many times, questions that are asked in the middle of a slide are answered by the end of the slide.

To minimize issues with the flow of your presentation (and to limit interruptions), you can ask people to wait until you finish each slide, or pause once in a while to ask your audience "Are there any questions up to this point?"

6. *Stay on time*: If you have half an hour for a presentation, try to allocate about half of your presentation to content and the other half to questions and answers. This has the added benefit of gauging participants' interest in your presentation.

7. *Color management*: Colors can increase participants' interest, but incorrect use can create distraction or make the presentation unintelligible. As a rule of thumb, try to keep to cool colors for the background of your slides (such as blue and green), and warmer colors (such as orange and red) for objects in the foreground. Also, be consistent throughout your presentation: once a color scheme is chosen, keep it for the remainder of the pre-sentation unless there is a specific reason for shifting (i.e., calling attention to a specific element). There are many online tools that can be used to manage colors so choose ones that match best. Some of the sites that I use include LOLColors (http://www.lolcolors.com/) and Coolors (https://coolors.co/).

6.3.1.1 Slideshow Software

There are a lot of alternative slideshow software programs available on the market. Here are some of the most popular or interesting ones out there:

■ *PowerPoint* is by far the most widespread slideshow soft-ware. It is well known by the majority of users, is simple to use, and can be quickly set up for presentations. It also exports slides as PDF files, making it easy to share content in a secure format. It comes with many templates pre-installed, but due to the prevalence of PowerPoint, many audience members will be familiar with most of the standard templates so if you use them, you won't be winning any points for originality.

■ *Google Slides* (https://www.google.ca/slides/about/) is a powerful alternative to PowerPoint. Like PowerPoint, it lets you integrate most file formats into your presentation, and integrates well into the Google software suite. The main strength of Google Slides is its collaborative aspect: it allows multiple users to access and update slides simultaneously. If you are working in a collaborative online environment, Google Slides might be the alternative you need.

■ *Slidebean* (slidebean.com) is another of many alternatives available in the slideshow space. The main advantage is that it has an easier learning curve, and can get start-up users quickly set up with an effective presentation as it handles most of the visual elements (letting you concentrate on the content).

6.3.1.2 Slideshow Framework: The 10/20/30 Rule

The 10/20/30 rule of presentation was proposed by Guy Kawasaki, a Silicon Valley venture capitalist specialized in marketing. Having seen numerous pitches and presentations, he noticed that entrepreneurs were over-allocating on unimportant content and losing focus. Since then, he has been advocating the 10/20/30 Framework. The objective is for the presenter to synthesize his corporate presentation down to the essentials: "A PowerPoint presentation should have *ten slides*, last no more than *twenty minutes*, and contain *no font smaller than thirty points*."* A complete description of his approach is available on his blog.

---

* More information on the 10/20/30 rule is available at http://guykawasaki.com/the_102030_rule/ (Accessed January 3, 2017).

The 10 slides that he deems essential are

1. The problem: What is the problem you have identified?
2. Your solution: How are you fixing the problem?
3. Business model: What is your organization business model?
4. Underlying magic/technology: One slide dedicated to your technological advantage.
5. Marketing and sales: What is your market, and how do you expect to generate sales?
6. Competition: What is the current competition (technologies, companies)?
7. Team: Who are the key members, and their strengths?
8. Projections and milestones: What are the key milestones to your success?
9. Status and time line: What is the current status of your technology, and your time line?
10. Summary and call to action.

The information you collect during your market research will be essential to building slides five (marketing and sales) and six (competition).

Afterwards, the onus is on the presenter to fit all of his information into a 20 minute presentation. As mentioned by M. Kawasaki, "In a perfect world, you give your pitch in twenty minutes, and you have forty minutes left for discussion."

Finally, he recommends using 30 point fonts, as he believes that too much text means that you will be reading your text rather than presenting your presentation. It forces the speaker to know his content, and to synthesize on the essentials.

6.3.2 Visual Storytelling Software

Visual storytelling software is an alternative presentation platform that enables interactive presentations. Rather than being confined within the chronological order of a slideshow, presenters can adjust their presentation to the audiences' interest and questions.

Billed as a conversational presentation tool, visual story-telling is a variation of the traditional top-down presentation. The rationale behind these platforms is that by the time your presentation starts, the attendees will already have researched a lot of information on you and your concept from your website and web presence, and they may already know what their gaps in knowledge and their interests are. They might already know what information they need to make a decision. Rather than you being the keeper of information, it is now the attend-ees who direct the presentation flow by asking questions and expressing interests. The objective is to have a natural conver-sation, and focus the presentation on what the participants are interested to find out.

One of the leaders in this space is Prezi (prezi.com), which has the additional advantage of being usable offline using an app, and is available in Android and desktop versions.

6.3.3 Infographics

While classical methods of presenting data are well known and recognized by the community, there is limited visual appeal to them. Also, there is very little that is distinctive about them. One of tools to emerge in this space is the use of *Infographics* to present data.

An infographic is a graphical visual display of information and data, which is used so that the information can be assimi-lated quickly by the reader in a pleasant manner. Hence, once you have found or created interesting and reliable data, info-graphics can be a useful tool to produce engaging visuals and create interest.

The advantage of using infographics is that they show overall context in a clear and visually appealing way. At a glance, the viewer is able to easily understand what the document is about as well as seeing how each data point is related to the others. Infographics are appealing because it is easier for people to process images than it is to process text. Also, they are a modern way of presenting otherwise traditional data and are portable, as you can easily insert them into a business plan, a webpage, a physical printout for distribution, or even post them across social media platforms.

Combining elements of table presentations with graphical elements, infographics allow the user to communicate messages in an attractive way to the audience. But to do so, the user will have to move beyond the traditional software suite. There are a number of interesting platforms that have powerful capabilities to build infographics, such as Canva Infographic Maker (www.canva.com/create/infographics/) and Venngage (venngage.com).

The downsides of infographics are that the researcher is limited in the complexity of the information he can share, and the abusive use of infographics can lead to situations akin to "PowerPoint poisoning," that is, the overload of information presented in such a disjointed way that it bores the recipients and loses their interest. As such, try to limit the amount of text on your infographic, privileging images and statistics instead.

One of the other main issues with infographics is data accuracy. If you have one incorrect data point on your infographic, it will quickly impact the overall credibility of the other data and the infographic in general, so make sure your data is completely accurate. Also, try to limit the number of colors you use on a single infographic: too many colors make it difficult for readers to understand the information you are trying to share with them. Finally, while you might have a lot of great information to share, be careful not to overload on content: excessive information will impact user readability. Making users wait for the page to load or having to scroll down will reduce the overall effectiveness of your infographic. If you have too much content, consider splitting the information into two separate documents, rather than trying to cram everything in a single document.

As an example, I have included an infographic I prepared for Ancon Medical, a client active in the early cancer detection space. This is based on the information found in his business plan as well as different market research reports I prepared for the company. Hence, I was able to build an infographic that conveys information on the importance of early cancer detection as well as showcasing some of the key relevant market data we found that supported the company's vision (Figure 6.5).

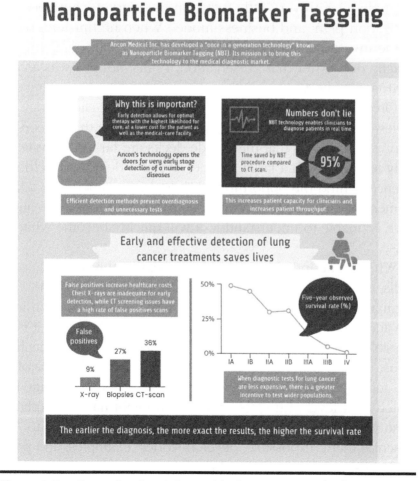

Figure 6.5   Example of an infographic for a company in the early cancer diagnostic space.

6.4 Closing Remarks: Marketing versus Technology

I remember attending a client's pitch a few years back. He had prepared his slide deck and felt quite confident about his presentation. He had invited me to attend to get the pulse from potential investors and to give him feedback on his presentation skills. He was quite proud of his technology. So proud, in fact, that his presentation focused almost exclusively on the technical innovation and prowess of this technology, with very little focus on the market, commercialization plan, and business model. When he finished, one of the investors candidly asked "What does your company exactly do?" After that presentation, I must admit I was as confused as he was.

On multiple occasions, I have been asked by clients to review and comment on their "pitch deck." One thing I constantly point out is that these slide decks are far too technically focused, with a large number of slides dedicated to the science behind the opportunity. While this is undeniably a key element of a company's unique selling proposition, if the audience is unable to understand it, it will quickly become white noise.

While some individuals possess exceptional research skills, creating paradigm-shifting technologies and redefining markets, they have to remember that not every person in the audience will possess the same level of skill. Hence, it is possible that the message may never reach the recipient as he is just unable to understand it. It is important to take note of the audiences' knowledge level and customize the message to match their level of understanding. You might have to explain what you feel are basic terms and concepts, so be ready to describe them. If possible, develop stories, examples, and parallels to simplify complex technologies.

Finally, in the initial meetings, some emphasis has to be given to the market opportunity that the technology provides, and how the company intends to approach the market to commercialize the technology. Later on, more time will be made available during due diligence to demonstrate the scientific and technical validity of the technology.

Chapter 7

Final Words

A few years back, as I was completing my graduate studies in communication, we were working on a case study related to crisis management. Hidden inside the case study was a simple throwaway line from one of the participants, who mentioned in passing that "Market Research creates our reality." To this day, I still think of this statement, as I believe that it candidly describes the importance that market research has on how the business world works.

Business leaders and decision makers need information if they want to make accurate decisions. As we mentioned in the introduction, market research is often performed when the cost of doing the research outweighs the cost of not doing it, and the dangers of making uninformed decisions run the risk of being costlier than doing nothing. Decision makers need that market data as part of their decision process.

From investors to future hires, everybody needs to know what the market is like. Investors need to know because they want to be sure that they are investing in the right place. Individuals need to know because they want to be sure that they are investing their most precious resources (their time) in the right place, not in a company that will go belly-up in a few years because it was building a product nobody needed.

Everybody in the business ecosystem is looking for the same thing, the answer to the question "Does the market exist?"

And its market research that brings this world to life.

Starting from an approximation that there is a market, somewhere out there, the market research you will be building with the tools you have gathered from this book will help to shape that story, find out just how many people need your product, and what exactly they need. From laboratory to market, the journey to learning your market can be iterative, and has to be ongoing, continually measuring your competitors' innovation through competitive intelligence activities to reaching out to your customers through voice of customer activities.

So where does that leave us?

As technology grows at a furious pace, so does our little corner of the market research world. Many of the tools and techniques that were common 10 years ago, are now classified as "traditional," still useful, but slowly being replaced by new technologies that revolutionize many of the approaches that were all too common a few years ago.

While attempting to be contemporary, I have reviewed and suggested a number of software programs and platforms to help in your market research, analysis, and classification. As you are reading this, some of the tools might already be defunct, replaced by newer and faster alternatives. Case in point, during the editing of this book, I found two platforms that were no longer available between the time I wrote the first manuscript and its final review before submitting it, three months later. Such is the curse of living in such a fast-paced world. Today's given is tomorrow's memory.

Tomorrow's market research effort will most likely depend more on passive secondary research tools, as these become more efficient and include learning algorithms to generate better searches and pre-analyze data. As technology improves, and people's knowledge of information technologies increases, market research will most likely shift to include more web-based online components, such as online focus groups and online interviews, slowly replacing traditional in-person focus groups and telephone interviews. Data that will be collected will increasingly be cross-checked with other databases: big data, a term that is overused, will probably take a more concrete form around consolidating data across organizations for more in-depth and effective research. Finally, platforms akin to IBM Watson that consolidate, analyze, and visualize data are most likely going to play a bigger role in how we interact with our data, and how we choose to share it.

Social media is also likely to change. Current monetization models that reward speed and quantity of information rather than accuracy create a pervasive world where good, useful information is drowned in the noise (both unintentionally and intentionally by organizations purposefully trying to generate noise to cover the right information). Social media can be an invaluable tool to gain insight, but triangulation is crucial to validate the information you gather from it.

Nonetheless, however technologies change, market research will continue to play a role in how organizations shape their interaction with their clients, their stakeholders, and their competitors.

It is my hope that this book gives you the tools you need to understand and share your story.

Bibliography and Further Reading

Anderson, C. 2010. Presenting and evaluating qualitative research. *American Journal of Pharmaceutical Education*. Vol. 74: (8) Article 141.

Camilli, G. and Shepard, L. A. 1994. *MMSS: Methods for Identifying Biased Test Items*. Thousand Oaks, CA: Sage.

Dariman, T. 2007. Exubera inhaled insulin discontinued. Diabetes Self-Management. Available at: https://www.diabetesselfman-agement.com/blog/exubera-inhaled-insulin-discontinued/ (Accessed December 25, 2016).

Edmunds, H. 1999. *The Focus Group Research Handbook* (American Marketing Association). Lincolnwood, IL: NTC Business Books.

Esponda, F. and Guerrero, V. 2009. Surveys with negative questions for sensitive items. *Statistics & Probability Letters*. Vol. 79(24): 2456–2461.

Few, S. Perceptual Edge. 2005. Effectively Communicating Number: Selecting the Best Means and Manner of Display (White paper).

Gordon, R. L. 1969. *Interviewing: Strategy, Techniques and Tactics*. Homewood, IL: The Dorsey Press.

Hedin, H., Hirvensalo, I. and Vaarnas, M. 2014. *The Handbook of Market Intelligence: Understand, Compete and Grow in Global Markets*. Chichester, UK: John Wiley & Sons.

Heinemann, L. 2008. The failure of exubera: Are we beating a dead horse? *Journal of Diabetes Science and Technology* (Online). Vol. 2(3): 518–529.

Holiday, R. 2013. *Trust Me, I'm Lying: Confessions of a Media Manipulator.* New York: Penguin Group.

Huebner, M., Werner, V. and le Cessie, S. 2016. A systematic approach to initial data analysis is good research practice. *Journal of Thoracic and Cardiovascular Surgery.* Vol. 151(1): 25–27.

Kelly, D. and Rupert, E. 2009. Professional emotions and persuasion: Tapping non-rational drivers in health-care market research. *Journal of Medical Marketing: Device, Diagnostic and Pharmaceutical Marketing.* Vol. 9(1): 3–9.

Kenichi Ohmae, British Library, People—https://www.bl.uk/peoplc/kenichi-ohmae (Accessed March 27, 2017).

Miles, M. and Huberman, M. 1994. *Qualitative Data Analysis: An Expanded Sourcebook* (Second Edition). Thousand Oaks, CA: Sage.

Mooi, E. and Sarstedt, M. 2014. *A Concise Guide to Market Research: The Process, Data and Methods Using IBMS SPSS Statistics* (Second Edition). Berlin: Springer.

Pit, S. W., Vo, T. and Pyakurel, S. 2014. The effectiveness of recruitment strategies on general practitioner's survey response rates: A systematic review. *BMC Medical Research Methodology.* doi:10.1186/1471-2288-14-76.

Rothwell, K. 2008. Ethics: The limits of intelligence gathering. *Competitive Intelligence Magazine.* Vol. 11(2): 34–35.

Sue, V. M. and Ritter, L. A. 2012. *Conducting Online Surveys* (2nd Edition). Thousand Oaks, CA: Sage.

Vach, W. 2013. Transformation of covariates. In *Regression Models as a Tool in Medical Research* (264–273). Boca Raton, FL: Taylor & Francis Group.

Verdinelli, S. and Scagnoli, N. 2013. Data display in qualitative research. *International Journal of Qualitative Methods.* Vol. 2013(12): 359–381.

Appendix: The Market Research Checklist

Sometimes, you want your research to go beyond the standard question of: "What's my market?" To help during the brainstorming and planning phase, here is a short list of question that can be useful to orient your research efforts. Questions break down into three distinct categories: your corporation (the market, the product), your consumers, and your competitors.

Your Corporation

- What is the size of the market?
- Is the market growing, steady or declining? At what rate?
- Is it an established market or is it an emerging market?
- Are there any changes in the environment that may affect your product such as government legislation, regulatory changes, taxation, or more?
- Is your product a disruptive innovation or an incremental innovation?
- What is the best way to promote the product?
- What should your commercialization strategy be?
- What should your pricing strategy be?

Your Customers

- Who is the purchaser of the product/service? Who is the decision maker?
- What is their purchasing process? How can you reach them?
- Who uses the product/service?
- Is the purchaser different then the end user? How do these two stakeholders interact?
- What is the product/service replacing?
- What is the user presently using to fill their current need?
- What do potential customers/end users think of the product/service?
- What is the profile of your customers (location, age, gender, income level, etc.)?
- What are their needs? What need does the product specifically target?
- What are the customer service and retention strategies that are needed?

Your Competitors

- What are the products that compete directly with your product? Indirectly?
- Who are the main competitors and what share of the market do they have?
  - What are their strengths and weaknesses?
  - Where do they have a competitive advantage?
  - What is their pricing strategy?
  - What is their branding strategy or Unique Selling Proposition?
- How can you differentiate your product from competing products?
- How is the market shaped?
  - What should your market segment be?
  - Are you positioning yourself correctly?
- How do customers compare you with your competitors?

Index

For Product Safety Concerns and Information please contact our EU
representative GPSR@taylorandfrancis.com Taylor & Francis Verlag GmbH,
Kaufingerstraße 24, 80331 München, Germany

Printed and bound by CPI Group (UK) Ltd, Croydon, CR0 4YY
08/05/2025
01864489-0003